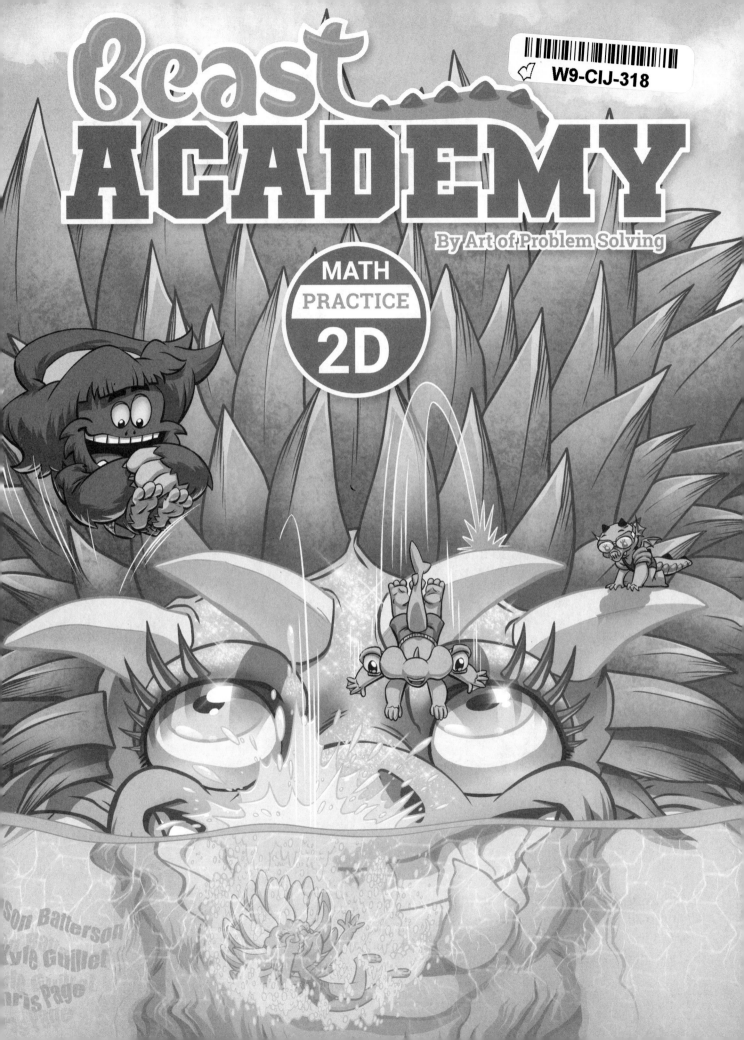

Published by: AoPS Incorporated
                       10865 Rancho Bernardo Rd Ste 100
                       San Diego, CA 92127-2102
                       info@BeastAcademy.com

ISBN: 978-1-934124-37-6

Beast Academy is a registered trademark of AoPS Incorporated.

Written by Jason Batterson, Kyle Guillet, and Chris Page
Book Design by Lisa T. Phan
Illustrations by Erich Owen
Grayscales by Greta Selman

Visit the Beast Academy website at BeastAcademy.com.
Visit the Art of Problem Solving website at artofproblemsolving.com.
Printed in the United States of America.
First Printing 2019.

Contents:

This is Practice Book 2D in a four-book series.

2A
• Place Value
• Comparing
• Addition

2B
• Subtraction
• Expressions
• Problem Solving

2C
• Measurement
• Strategies (+&−)
• Odds & Evens

2D
• Big Numbers
• Algorithms (+&−)
• Problem Solving

For more resources and information, visit BeastAcademy.com.

This is Beast Academy Practice Book 2D.

MATH
PRACTICE
2D

Each chapter of this Practice book corresponds to a chapter from Beast Academy Guide 2D.

MATH
GUIDE
2D

The first page of each chapter includes a recommended sequence for the Guide and Practice books.

You may also read the entire chapter in the Guide before beginning the Practice chapter.

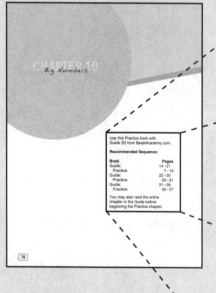

Use this Practice book with Guide 2D from BeastAcademy.com.

Recommended Sequence:

Book	Pages
Guide:	14–21
Practice:	7–19
Guide:	22–30
Practice:	20–31
Guide:	31–39
Practice:	32–37

You may also read the entire chapter in the Guide before beginning the Practice chapter.

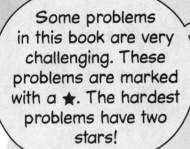

Some problems in this book are very challenging. These problems are marked with a ★. The hardest problems have two stars!

Every problem marked with a ★ has a *hint!*

Hints for the starred problems begin on page 106.

42   Guide Pages: 39-43

Some pages direct you to related pages from the Guide.

None of the problems in this book require the use of a calculator.

Solutions are in the back, starting on page 110.

A complete explanation is given for every problem!

CHAPTER 10
Big Numbers

Use this Practice book with Guide 2D from BeastAcademy.com.

Recommended Sequence:

You may also read the entire chapter in the Guide before beginning the Practice chapter.

A digit's location in a number is called its **place value**. The three-digit number below has a hundreds place, a tens place, and a ones place.

hundreds
tens
ones

987

987 has hundreds digit 9, tens digit 8, and ones digit 7.

PRACTICE | Solve each problem below.

1. Circle every number below with tens digit 8.

   18   81   118   188   818   881

2. Circle every number below whose ones digit is the same as its hundreds digit.

   737   337   733   373   773   377

3. Circle every number below whose ones digit is greater than its tens digit.

   92   229   292   922   992   29

4. Arrange the digits 0, 2, and 4 to create a 3-digit number that has a 4 in the tens place.

   4. _____

5. Arrange the digits 5, 6, and 7 to create a 3-digit number so that:
   • The ones digit is greater than the tens digit.
   • The ones digit is less than the hundreds digit.

   5. _____

The largest three-digit number is 999.
The whole number after 999 is one thousand: 1,000.
The **thousands** place is to the left of the hundreds place.

The whole number after 9,999 is ten thousand: 10,000.
The **ten-thousands** place is to the left of the thousands place.

The whole number after 99,999 is one hundred thousand: 100,000.
The **hundred-thousands** place is to the left of the ten-thousands place.

hundred thousands
ten thousands
thousands
hundreds
tens
ones

100,000

We usually write a comma between the thousands place and the hundreds place.

This makes big numbers easier to read.

PRACTICE | Solve each problem below.

6. Circle the ten-thousands digit in each number below.

87,209   15,426   378,921   356,278

7. Circle the thousands digit in each number below.

42,809   75,021   529,481   123,456

8. Circle the numbers below that have hundred-thousands digit 7.

78,786   725,325   177,771   707,070

We can count by thousands from one thousand (1,000) to nine hundred ninety-nine thousand (999,000).

One thousand more than 999,000 is one thousand thousands! We call this one *million*: 1,000,000.

The *millions* place is to the left of the hundred-thousands place. To the left of the millions place is the *ten-millions* place, followed by the *hundred-millions* place.

hundred millions
ten millions
millions
hundred thousands
ten thousands
thousands
hundreds
tens
ones

100,000,000

We use commas to separate digits into groups of 3.

PRACTICE | Solve each problem below.

9. Circle the ten-millions digit in each number below.

174,870,932   89,647,150   706,813,924   30,458,972

10. Circle every number below whose millions digit is the same as its ten-thousands digit.

808,080,808   123,123,123   234,543,210   567,765,567

11. Circle every number below whose tens digit is greater than its ten-millions digit.

123,456,789   978,675,645   232,121,434   624,910,538

When reading big numbers like the ones below, we read the number of millions, then the number of thousands, then the rest.

12,345,060   3,000,003   900,090,000

12,345,060 has 12 millions, 345 thousands, and 60. So, we read 12,345,060 as "twelve million, three hundred forty-five thousand, sixty."

3,000,003 has 3 millions, no thousands, and 3. So, we read 3,000,003 as "three million three."

900,090,000 has 900 millions and 90 thousands. So, we read 900,090,000 as "nine hundred million, ninety thousand."

If you can read three-digit numbers, you can read bigger numbers, too.

PRACTICE | Draw a line to connect each number on the left with the matching number on the right.

12. 5,050,500

Five million, five thousand five

13. 500,050,000

Five hundred million, fifty thousand

14. 550,000

Five million five

15. 5,005,005

Fifty thousand, five hundred

16. 50,500

Five hundred fifty thousand

17. 5,000,005

Five million, fifty thousand, five hundred

When writing a number in words, we sometimes include commas to make it easier to read.

PRACTICE | Use digits to write each number below.

18. Three hundred forty-five thousand, nine hundred eighty-six.

18. _____

19. Seventy-eight million, three hundred forty thousand.

19. _____

20. Four hundred six million, nine hundred seventy-two.

20. _____

21. Thirty million, seven thousand.

21. _____

22. Seven hundred million, twenty-four thousand, six hundred thirty-eight.

22. _____

23. Fifty-nine million fifty-nine.

23. _____

24. Two hundred twelve thousand, two hundred twelve.

24. _____

25. Eight hundred two million, two thousand two.

25. _____

BIG NUMBERS

In a **Numbercross** puzzle, the goal is to fill all of the blanks in a grid with digits so that every number that is written in words appears in the grid. Numbers in the grid can be read either left-to-right or top-to-bottom.

EXAMPLE

Use the clues below to complete the Numbercross puzzle on the right.

Two hundred two
Two thousand five hundred
Five thousand five
Twenty-five thousand twenty
Twenty-five thousand two hundred
Two hundred thousand, two hundred fifty

First, we write each number using digits.

202
2,500
5,005
25,020
25,200
200,250

We can arrange these numbers as shown so that every number appears in the grid.

2	0	2			2
5		5	0	0	5
0		2			0
2	0	0	2	5	0
0		0			

PRACTICE | Use the clues to complete each Numbercross puzzle below.

26.
Nine hundred nine
Nine hundred ninety
Nine thousand ninety-nine
Ninety-nine thousand ninety

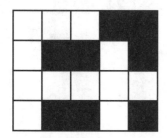

27.
Seven hundred seventy
Seven thousand seventy
Seven thousand seven hundred
Seventy-seven thousand seven

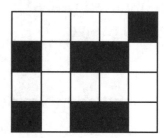

PRACTICE | Use the clues to complete each Numbercross puzzle below.

28. One thousand two
Nine thousand seventy
Ten thousand, six hundred four
Twenty thousand, five hundred three
Eighty-four thousand three
Seven hundred six thousand five

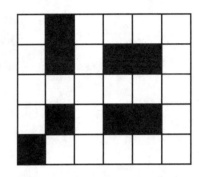

29. Three hundred seventy-five
Three thousand seventy-five
Fifty thousand, three hundred seven
Seventy-five thousand three
Three hundred seventy thousand, five hundred five
Five hundred thirty thousand, three hundred

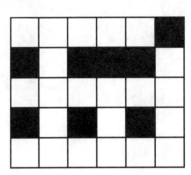

30. ★ Three hundred three
Four thousand four
Thirty thousand three hundred
Fifty thousand, three hundred thirty
Fifty-five thousand fifty
Six hundred thousand, three hundred four

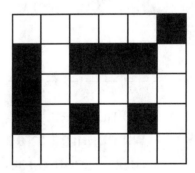

31. ★ Four hundred forty
Four thousand four
Four thousand four hundred
Forty thousand forty
Forty thousand, four hundred four
Forty-four thousand, four hundred

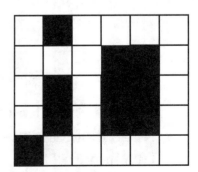

A **Numberlap** puzzle has a row of blanks, with numbers written in words beneath those blanks. The goal is to fill the blanks with digits so that each number appears in the given row.

EXAMPLE | Solve the Numberlap puzzle below.

— — — — — — —

Eight hundred ninety
One thousand eighty-nine
Nine thousand six

We fill the blanks with digits so that each number appears in the row.

1 0 8 9 0 0 6

Eight hundred ninety   1 0 **8 9 0** 0 6
One thousand eighty-nine   **1 0 8 9** 0 0 6
Nine thousand six   1 0 8 **9 0 0 6**

PRACTICE | Solve each Numberlap puzzle below.

32. — — — — — — —

One hundred two
Two hundred three
Three hundred four

33. — — — — — — —

Five thousand seventy
Five thousand, five hundred seven
Six thousand fifty-five

34. — — — — — — —

One thousand twenty-three
Three thousand twenty-one
Two hundred ten

35. — — — — — — —

One hundred thirty-one
One thousand thirty-one
Three thousand, one hundred thirteen

PRACTICE | Solve each Numberlap puzzle below.

36. __ __ __ __ __ __ __ __

Twenty thousand twelve
One thousand two hundred
Two hundred two

37. __ __ __ __ __ __ __ __ __

Five hundred six
Four thousand fifty
Five thousand forty
Six thousand fifty

38.
★ __ __ __ __ __ __ __ __ __ __

Seven thousand, five hundred seven
Five thousand seventy-five
Five thousand seventy-seven
Seven thousand seventy-seven

39.
★
★ __ __ __ __ __ __ __ __ __ __ __

Nine thousand, four hundred forty-nine
Four thousand forty-four
Four thousand, four hundred ninety-four
Forty-nine thousand forty-four

Consecutive numbers come one right after another.
For example, 23, 24, and 25 are consecutive numbers.

EXAMPLE | Fill the blanks below to create a list of consecutive numbers.

_____ 9,997 _____ _____ _____

One less than 9,997 is 9,996. Counting up from 9,997 we have 9,998, then 9,999. Finally, one more than 9,999 is 10,000.

We fill the blanks as shown.

9,996   9,997   **9,998**   9,999   **10,000**

PRACTICE | Complete each list of consecutive numbers below.

40. __995__ __996__ _____ _____ _____ _____ __1,001__

41. __179,997__ _____ _____ _____ __180,001__

42. _____ _____ __84,997__ __84,998__ _____ _____

43. __1,099,998__ __1,099,999__ _____ _____

44. __3,899,997__ __3,899,998__ _____ _____

45. _____ _____ _____ __100,002__ __100,003__

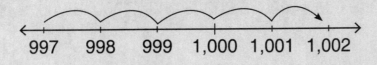

EXAMPLE | What is 997+5?

We can count up by 5 from 997.

997   998   999   1,000   1,001   1,002

So, 997+5 = **1,002**.

– *or* –

To add 5 to 997, we add 3 then add 2.
997+3 = 1,000. Then, 1,000+2 = **1,002**.

Sometimes, adding or subtracting even a small amount can be tricky.

PRACTICE | Fill in the blanks below.

46.   3 *more* than 99,999 is _____ .

47.   4 *less* than 10,001 is _____ .

48.   1,998+3 = _____

49.   140,002−5 = _____

50.   99,996+7 = _____

51.   1,000,004−6 = _____

52.   31,992+11 = _____

53.   45,002−9 = _____

54.   49,996+12 = _____

55. ★   103,002−4 = _____

56.   9,099,099+6 = _____

57. ★   1,011,001−7 = _____

We can count by 10's, 100's, and 1,000's.

10's: 10 20 30 40 50 60 70 80 90 100 110 120 130 140 150 160 170

100's: 100 200 300 400 500 600 700 800 900 1,000 1,100 1,200 1,300

1,000's: 1,000 2,000 3,000 4,000 5,000 6,000 7,000 8,000 9,000 10,000

EXAMPLE | Count by 100's to complete the list of numbers below.

    __9,700__   __9,800__   _____   _____   _____

We keep adding 100 to fill the next three blanks as shown.

    __9,700__   __9,800__   **9,900**   **10,000**   **10,100**

> We can count by numbers other than 1.

PRACTICE | Count by the number given to fill the blanks in each list below.

58. **10's:** __950__   __960__   _____   _____   _____

59. **1,000's:** __196,000__   __197,000__   _____   _____   _____

60. **100's:** __39,600__   __39,700__   _____   _____   _____

61. **10's:** __109,970__   __109,980__   _____   _____   _____

62. **100's:** __89,800__   __89,900__   _____   _____   _____

63. **1,000's:** __1,098,000__   __1,099,000__   _____   _____   _____

PRACTICE | Solve each problem below.

64. Write each number described below. The first blank has been filled for you.

8 hundreds = ___800___   10 hundreds = _____

13 hundreds = _____   17 hundreds = _____

20 hundreds = _____   100 hundreds = _____

65. Write each number described below.

5 tens = _____   9 tens = _____

10 tens = _____   13 tens = _____

40 tens = _____   555 tens = _____

66. Write each number described below.

7 thousands = _____   70 thousands = _____

700 thousands = _____   4,000 thousands = _____

67. ★ Write each number described below.

100,000 tens = _____

100,000 hundreds = _____

100,000 thousands = _____

EXAMPLE | Add: 123,000+5,000
123,000+50,000
123,000+500,000

123,000+5,000: 123 thousands plus 5 thousands is
123+5 = 128 thousands, or **128,000**.

123,000+50,000: 123 thousands plus 50 thousands is
123+50 = 173 thousands, or **173,000**.

123,000+500,000: 123 thousands plus 500 thousands is
123+500 = 623 thousands, or **623,000**.

> Adding 5,000 increases the thousands digit by 5.

> Adding 50,000 increases the ten-thousands digit by 5.

> Adding 500,000 increases the hundred-thousands digit by 5.

PRACTICE | Solve each addition problem below.

68. 12,000+1,000 = _____

69. 41,000+6,000 = _____

70. 333,000+10,000 = _____

71. 35,000+40,000 = _____

72. 650,000+300,000 = _____

73. 4,567,000+20,000 = _____

74. 713,975+80,000 = _____

75. 92,756+6,000 = _____

76. ★ 154,103+30,070 = _____

77. ★ 375,232+500,600 = _____

We can subtract by place value, too!

PRACTICE | Solve each subtraction problem below.

78. 35,000 – 1,000 = _____

79. 247,000 – 5,000 = _____

80. 290,000 – 10,000 = _____

81. 450,000 – 30,000 = _____

82. 980,000 – 600,000 = _____

83. 1,290,000 – 70,000 = _____

84. 239,654 – 5,000 = _____

85. 78,987 – 60,000 = _____

86. ★ 756,492 – 200,300 = _____

87. ★ 796,562 – 80,020 = _____

88. ★ What is 20 thousands minus 20 hundreds minus 20 tens?

88. _____

89. ★ What do you get if you subtract 5 tens, 5 thousands, and 5 millions from 9,090,900?

89. _____

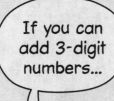

BIG NUMBERS

If you can add 3-digit numbers...

...you can add bigger numbers, too.

EXAMPLE | Add 156,000+18,000.

156 thousands plus 18 thousands is
156+18 = 174 thousands, or **174,000**.

EXAMPLE | Add 130,780+70,098.

We can turn this into two easier addition problems.

We start by adding the thousands.
130 thousands plus 70 thousands is
130+70 = 200 thousands, or 200,000.

Then, adding the rest, we get 780+98 = 878.

So, 130,780+70,098 = 200,000+878,
which is **200,878**.

PRACTICE | Fill in the blanks to solve each addition problem below.

90. $11,250+55,075 =$ ☐☐,0 0 0 + 3 2 5 = ☐☐,☐☐☐

91. $80,090+45,070 =$ ☐☐☐,0 0 0 + ☐☐ = ☐☐☐,☐☐☐

92. $246,392+98,208 =$ ☐☐☐,0 0 0 + ☐☐☐ = ☐☐☐,☐☐☐

93. ⭐ $43,480,278+35,070,300 =$ ☐☐,☐☐☐,☐☐☐

EXAMPLE | Add 16,700+19,800.

Adding the thousands gives us
16+19 = 35 thousands, or 35,000.

Adding the rest gives us
7+8 = 15 hundreds, or 1,500.

Finally, 35,000+1,500 = **36,500**.

It's easy to lose track if you aren't writing things down as you go.

PRACTICE | Fill in the blanks to solve each problem below.

94. 3,900+4,700 = $\boxed{7}\boxed{0}\boxed{0}\boxed{0}$ + $\boxed{},\boxed{}\boxed{}\boxed{}$ = $\boxed{},\boxed{}\boxed{}\boxed{}$

95. 800,000+300,000 = $\boxed{},\boxed{}\boxed{}\boxed{},\boxed{0}\boxed{0}\boxed{0}$

96. 30,400+50,700 = $\boxed{},\boxed{0}\boxed{0}\boxed{0}$ + $\boxed{},\boxed{}\boxed{}\boxed{}$ = $\boxed{},\boxed{}\boxed{}\boxed{}$

97. ★ 500,500+800,800 = $\boxed{},\boxed{}\boxed{}\boxed{},\boxed{}\boxed{}\boxed{}$

98. ★ 750,600+740,900 = $\boxed{},\boxed{}\boxed{}\boxed{},\boxed{}\boxed{}\boxed{}$

EXAMPLE

In a distant planetary system, planet Zort is 47,000,000 miles from the sun and planet Bort is 91,000,000 miles from the sun. When Zort and Bort are on opposite sides of the sun, how many miles apart are they?

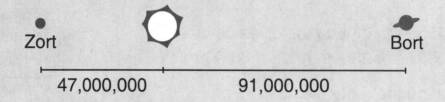

Zort

Bort

47,000,000   91,000,000

Adding the distances in millions gives $47 + 91 = 138$ million miles, or **138,000,000** miles.

PRACTICE | Answer each word problem below.

99. It takes 80,000 miles of cable to make one bridge connecting Beast Bay Island to Kowabunga Key. How many miles of cable are needed for two bridges?

99. _____

100. A mountain yeti has about 1,900,000 hairs. A snow yeti has about 500,000 more hairs than a mountain yeti. About how many hairs does a snow yeti have?

100. _____

101. Researchers estimate that there are 1,250,000 tons of sand at Blowfish Beach and 1,580,000 tons of sand at Coral Cove. Using these estimates, how many more tons of sand are there at Coral Cove than at Blowfish Beach?

101. _____

PRACTICE | Answer each word problem below.

102. What is the sum of the three **smallest** 5-digit numbers?

102. _____

103. There are 5,280 feet in 1 mile. How many feet are there in 2 miles?

103. _____

104. There are 86,400 seconds in 1 day, and there are 604,800 seconds in 1 week. How many seconds are there in 8 days?

104. _____

105. ★ What do you get when you add the **largest** number you can make with three 3's and three 0's to the **smallest** number you can make with three 3's and three 0's?

105. _____

106. ★ How much greater than 99,000 is 99,000,000?

106. _____

EXAMPLE | Order the numbers below from least to greatest.

345,000   1,212,000   91,900   91,500   800,000

We start by looking for the numbers with the fewest digits.

There are two 5-digit numbers: 91,900 and 91,500.
500 is less than 900, so 91,500 is less than 91,900.

There are two 6-digit numbers: 345,000 and 800,000.
345 is less than 800, so 345,000 is less than 800,000.

Finally, the 7-digit number 1,212,000 is greatest.
So, we order the numbers from least to greatest as shown.

91,500   91,900   345,000   800,000   1,212,000

> When comparing two whole numbers that don't have the same number of digits...
>
> ...the number with more digits is always greater.

PRACTICE | Fill each circle below with < or >.

107. 23,232 ◯ 32,323

108. 98,765 ◯ 456,789

109. 101,010 ◯ 98,989

110. 2,181,200 ◯ 2,179,900

111. 45,678 ◯ 45,876

112. 9,009,009 ◯ 10,001,001

113. 777,777 ◯ 88,888

114. 2,000,001 ◯ 1,999,998

PRACTICE | Solve each problem below.

115. Fill in the blanks to order the numbers below from *least* to *greatest*.

905,000   9,500,000   95,000   90,500,000   950,000

_____ < _____ < _____ < _____ < _____

116. Fill in the blanks to order the numbers below from *greatest* to *least*.

11,111   2,222,222   33,333   4,444,444   555,555

_____ > _____ > _____ > _____ > _____

117. Circle the *greatest* amount below.

6 ten-thousands   8 thousands   9 hundreds

7 hundred-thousands   5 thousands

118. Circle the *greatest* amount below.
★

5,500 thousands   5 millions   55,555 tens

550 thousands   5,555 hundreds

PRACTICE | Solve each problem below.

119. What is the largest four-digit number whose digits are all different?

119. _____

120. What is the smallest five-digit number that uses five different **even** digits?

120. _____

121. Fill all of the blanks below with the **same digit** to make a true statement.

$$4{,}123 < \boxed{}{,}000 < 5{,}\boxed{}123$$

122. ★ What is the smallest six-digit number where every digit is larger than the digit to its right?

122. _____

123. ★ Alex writes all of the four-digit numbers that can be made using the digits 1, 2, 3, and 4 once each. He lists his numbers from least to greatest. What is the 5th number in Alex's list?

123. _____

PRACTICE | Solve each problem below.

124. Ms. Q. writes 21089 on the board. If she erases the 8, she is left with the 4-digit number 2,109. What is the smallest 4-digit number Ms. Q. can leave after erasing **one** of the digits in 21089?

124. _____

21089

125. What is the smallest 4-digit number Ms. Q. can leave after erasing **one** of the digits in 50648?
★

125. _____

50648

126. What is the smallest 5-digit number Ms. Q. can leave after erasing **two** of the digits in 2745638?
★

126. _____

2745638

127. Ms. Q. writes the numbers from 1 to 15 back-to-back to create the number below. What is the smallest 12-digit number Ms. Q. can leave after erasing **nine** of these digits?
★

127. _____

123456789101112131415

EXAMPLE | An extra digit has been inserted in the statement below! Cross out the extra digit to make the statement true.

$$3\ 5\ 3\ 5\ 3 < 3\ 5\ 3\ 3$$

The left number has more digits than the right number. Since the left number must be smaller than the right number, we must cross out one of the left number's digits.

Only crossing out the digit below gives a true statement.

$$3\ \cancel{5}\ 3\ 5\ 3 < 3\ 5\ 3\ 3$$

Check: 3,353 < 3,533. ✓

PRACTICE | Cross out the given number of digits to make each statement true.

128. Cross out: 1 digit

$$1\ 2\ 3\ 4 < 1\ 2\ 4$$

129. Cross out: 1 digit

$$7\ 6\ 5\ 4\ 3\ 2 < 6\ 7\ 8\ 9\ 0$$

130. Cross out: 1 digit

$$2\ 4\ 2\ 4\ 2\ 4 < 2\ 3\ 4\ 5\ 6$$

PRACTICE | Cross out the given number of digits to make each statement true.

131. Cross out: 1 digit

$$4\ 5\ 5\ 4 < 5\ 4\ 5\ 4\ 5 < 5\ 4\ 5\ 4$$

132. Cross out: 2 digits

$$9\ 9\ 9\ 8 < 8\ 9\ 9\ 9\ 9\ 8 < 1\ 0\ 0\ 0\ 0$$

133. Cross out: 2 digits

$$4\ 5\ 6\ 7 < 3\ 4\ 5\ 7 < 1\ 1\ 1\ 1$$

134. Cross out: 2 digits

$$6\ 8\ 2\ 4 < 4\ 6\ 8\ 2 < 2\ 4\ 6\ 8$$

135. Cross out: 3 digits
★

$$4\ 2\ 3\ 1 < 2\ 3\ 4\ 1 < 2\ 4\ 1\ 3 < 2\ 4\ 3$$

EXAMPLE | Is 4,555 closer to 4,000 or to 5,000?

The number halfway between 4,000 and 5,000 is 4,500, and 4,555 is more than 4,500.

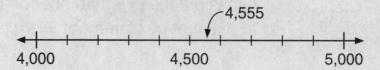
4,555

4,000   4,500   5,000

So, 4,555 is closer to **5,000** than it is to 4,000.

Sometimes we use a nearby number to describe a large amount.

For example, we might say that the moon is 240,000 miles away, even though it's not **exactly** 240,000 miles away.

PRACTICE | Solve each problem below.

136. What number is halfway between 30,000 and 40,000?

136. _____

137. Is 34,356 closer to 30,000 or to 40,000?

137. _____

138. Is 652,476 closer to 600,000 or to 700,000?

138. _____

139. The population of Dragonia is 283,791. For each pair of numbers below, circle the one that is closest to the population of Dragonia.

283,700 or 283,800

280,000 or 290,000

200,000 or 300,000

PRACTICE | Solve each problem below.

140. Fill the empty blank in each number below with a digit so that each number is as close as possible to 5,468.

5,4☐0   5,☐00   ☐,000

141. Fill the empty blank in each number below with a digit so that each number is as close as possible to 87,290.

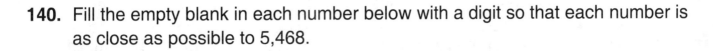

87,☐00   8☐,000   ☐0,000

142. Fill the empty blanks in each number below with digits so that each number is as close as possible to 45,827.

☐☐,☐00   ☐☐,000   ☐0,000

143. ★ Fill the empty blanks in each number below with digits so that each number is as close as possible to 764,971.

☐☐0,000   ☐☐☐,000   ☐☐☐,☐00

When we don't need to know an exact amount, sometimes we **estimate**.

A good estimate is easy to compute in your head **and** close to the right amount.

EXAMPLE | Estimate 15,776+29,283.

Since 15,776 is close to 16,000 and 29,283 is close to 29,000, we estimate that 15,776+29,283 is about 16,000+29,000 = **45,000**.

This estimate is easy to compute in our head. Since we didn't change either number by much, it's also very close to the exact answer, 45,059.

We could have used 16,000+30,000 = **46,000**, which is a little easier to compute, but not quite as close.

A sum like 15,780+29,280 is not much easier to compute than the original sum, so it is not very useful for estimation.

A sum like 20,000+30,000 is very easy to compute in our head, but not as close to the exact answer, so it is not as good an estimate.

The amount you are off by should be small compared to your estimate.

PRACTICE | Solve each problem below.

144. The population of Aytown is 712,743. The population of Beeville is 406,459. Circle the sum below that is **best** used to estimate the total number of people living in Aytown and Beeville.

712,740+406,460   712,700+406,500   700,000+400,000

145. Broadzilla weighs 142,862 pounds. Queen Kong weighs 128,541 pounds. Circle the sum below that is **best** used to estimate the combined weight of Broadzilla and Queen Kong.

142,900+128,500   140,000+130,000   100,000+100,000

PRACTICE | Solve each problem below.

146. Circle the number below that is closest to 22,538+38,902.

20,000   40,000   60,000   80,000   100,000

147. Circle the number below that is closest to 185,057+476,336.

66,000   560,000   660,000   760,000   1,000,000

148. Circle the letter that marks 4,097,865+1,024,196 on the number line below.

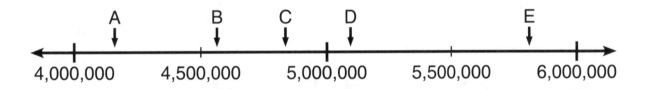

PRACTICE | Fill the blanks below with < or > **without** finding the actual sums.

149. 697+691 ◯ 1,400

150. 30,052+40,190 ◯ 70,000

151. 7,950+4,925 ◯ 13,000

152. 200,000 ◯ 128,128+69,214

153. ★ 486+679 ◯ 317+926

154. ★ 5,778+8,976 ◯ 3,123+12,430

PRACTICE | Solve each problem below.

155. What is the largest whole number that is closer to 10,000 than it is to 20,000?

155. _____

156. What is thirty-three thousand thirty-three plus forty-four million forty-four?

156. _____

157. ★ What do you get when you subtract the smallest number below from the largest number below?

111,111   3,333,333   555,555   777,777   99,999

157. _____

158. ★ What is the sum of the three *largest* 5-digit numbers?

158. _____

159. ★ Alex writes the number 6,677. Then, he switches the thousands digit and the ones digit. How much greater is the result?

159. _____

PRACTICE | Solve each problem below.

160. Circle the *two* numbers below that have a sum of 38,204.

10,927   17,357   20,847   24,437   29,187

161. Use the digits 4, 5, 5, and 6 to make a four-digit
★ number that is as close to 5,000 as possible.

161. _____

162. The number 9,990 has 9 thousands, 9 hundreds,
★ and 9 tens. How would you write the number that
has 10 thousands, 10 hundreds, and 10 tens?

162. _____

163. What is the smallest 7-digit number whose digits
★ add up to 20?
★

163. _____

CHAPTER 11
Algorithms

Use this Practice book with
Guide 2D from BeastAcademy.com.

Recommended Sequence:

Book	Pages:
Guide:	42-53
Practice:	39-47
Guide:	54-63
Practice:	48-65
Guide:	64-73
Practice:	66-69

You may also read the entire
chapter in the Guide before
beginning the Practice chapter.

An **algorithm** is a step-by-step process you can use to do something.

PRACTICE | Use the steps below to continue each pattern.

Step 1: Choose any whole number.

Step 2: If your number is even, find half of your number. If your number is odd, add three copies of your number, then add 1. Write the result.

Step 3: Repeat step 2 with your new number.

1. Fill the blanks to continue the list below using the algorithm above.

__6__ , __3__ , __10__ , _____, _____, _____, _____, _____, _____, _____

2. Fill the blanks to continue the list below using the algorithm above.

__7__ , __22__ , _____, _____, _____, _____, _____, _____, _____, _____

3. Fill the blanks to continue the list below using the algorithm above.

__4__ , _____, _____, _____, _____, _____, _____, _____, _____, _____

Math beasts think that no matter what number you start with, you will eventually get to 1, but no one knows for sure. If you start with 97, it will take 118 steps to get to 1!

> Stacking addition to line up the place values can help us organize our work.

EXAMPLE | Compute 85+67.

We can add the tens and the ones separately.

$$\overset{8 \atop \text{tens}}{} + \overset{6 \atop \text{tens}}{} = \overset{14 \atop \text{tens}}{}$$

$$85 + 67 = 140 + 12 = 152$$

$$\underset{\text{ones}}{5} + \underset{\text{ones}}{7} = \underset{\text{ones}}{12}$$

So, 85+67 = 140+12 = **152**.

— *or* —

We can stack the numbers on top of each other to line up the place values, then add the ones and the tens separately.

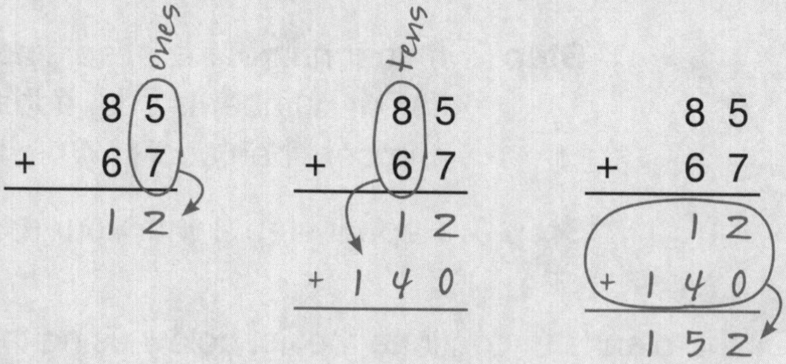

```
       ones                tens
   8  5               8  5            8  5
+  6  7            +  6  7         +  6  7
───────           ───────         ───────
   1  2               1  2            1  2
                  + 1  4  0       + 1  4  0
                  ─────────       ─────────
                                    1  5  2
```

So, 85+67 = 12+140 = **152**.

PRACTICE | Find each sum by stacking as shown above.

4.
```
    3 7
  + 8 6
  ─────

+
```

5.
```
    5 4
  + 3 9
  ─────

+
```

6.
```
      8 9
    + 5 3
    ─────

+
```

A Common Addition Algorithm:

Below, we show how you can use an algorithm to add 85+67.

Step 1: Stack the numbers you are adding, lining up the digits by place value.

$$\begin{array}{r} 8\ 5 \\ +\ 6\ 7 \\ \hline \end{array}$$

Step 2: Add the ones. 5+7 = 12 ones, which is 1 ten and 2 ones.

Put the 2 at the **bottom** of the ones column, and the 1 at the **top** of the tens column.

$$\begin{array}{r} \scriptstyle 1 \\ 8\ 5 \\ +\ 6\ 7 \\ \hline 2 \end{array}$$

Step 3: Add the tens. 1+8+6 = 15 tens, which is 1 hundred and 5 tens.

Put the 5 at the **bottom** of the tens column. Since there are no more hundreds, we place the 1 at the bottom of the hundreds column.

$$\begin{array}{r} \scriptstyle 1 \\ 8\ 5 \\ +\ 6\ 7 \\ \hline 1\ 5\ 2 \end{array}$$

So, 85+67 = **152**.

PRACTICE | Find each sum using the algorithm shown above.

7.
$$\begin{array}{r} 3\ 7 \\ +\ 8\ 6 \\ \hline \end{array}$$

8.
$$\begin{array}{r} 5\ 4 \\ +\ 3\ 9 \\ \hline \end{array}$$

9.
$$\begin{array}{r} 8\ 9 \\ +\ 5\ 3 \\ \hline \end{array}$$

ALGORITHMS
Stacking

EXAMPLE | Compute 448+275.

We stack the addition so that the place values line up.
Then, we add each place value, working from right to left.

We work from right to left so that we can regroup into larger place values as we go.

Add the
ones.

```
   1
  4 4 8
+ 2 7 5
───────
      3
```

Add the
tens.

```
  1 1
  4 4 8
+ 2 7 5
───────
    2 3
```

Add the
hundreds.

```
  1 1
  4 4 8
+ 2 7 5
───────
  7 2 3
```

So, 448+275 = **723**.

PRACTICE | Compute each sum using the algorithm shown above.

10.
```
  2 4 6
+ 5 7 9
───────
```

11.
```
  1 5 8
+ 3 4 7
───────
```

12.
```
  3 5 8
+ 9 3 4
───────
```

13.
```
  4 7 3
+ 6 8 4
───────
```

14.
```
  2 5 6
+ 4 5 9
───────
```

15.
```
  7 8 5
+ 8 1 6
───────
```

PRACTICE | Fill in the blanks to compute each sum below.

16.
$$
\begin{array}{r}
8\ 7\ 6 \\
+\ 5\ 4\ 3 \\
\hline
\square\,,\square\ \square\ \square
\end{array}
$$

17.
$$
\begin{array}{r}
2\ 2\ 7 \\
+\ 4\ 1\ 4 \\
\hline
\square\ \square\ \square
\end{array}
$$

18.
$$
\begin{array}{r}
2\ 4\ 6 \\
+\ 3\ 6\ 9 \\
\hline
\square\ \square\ \square
\end{array}
$$

19.
$$
\begin{array}{r}
4\ 4\ 4 \\
+\ 7\ 7\ 7 \\
\hline
\square\,,\square\ \square\ \square
\end{array}
$$

20.
$$
\begin{array}{r}
7\ 6\ 5 \\
+\ 1\ 4\ 7 \\
\hline
\square\ \square\ \square
\end{array}
$$

21.
$$
\begin{array}{r}
8\ 3\ 3 \\
+\ 3\ 3\ 8 \\
\hline
\square\,,\square\ \square\ \square
\end{array}
$$

22.
$$
\begin{array}{r}
3\,,6\ 0\ 9 \\
+\ \ \ 8\ 7\ 4 \\
\hline
\square\,,\square\ \square\ \square
\end{array}
$$

23.
$$
\begin{array}{r}
5\,,0\ 8\ 1 \\
+\ 4\,,9\ 8\ 3 \\
\hline
\square\ \square\,,\square\ \square\ \square
\end{array}
$$

Try these ridiculous sums!

PRACTICE | Compute each sum below.

24.
```
      □   □  □         □
     8 , 6 7 5 , 3 0 9
   + 8 , 6 7 5 , 3 0 9
   □ □ , □ □ □ , □ □ □
```

25.
```
     □ □ □   □ □ □   □ □
    1 2 3 , 4 5 6 , 7 8 9
  + 9 8 7 , 6 5 4 , 3 2 1
  □ , □ □ □ , □ □ □ , □ □ □
```

26.
```
   □ □   □ □ □ □   □
  6 5 , 6 5 6 , 5 6 5 , 6 5 6
+      7 , 4 3 7 , 4 3 7 , 4 3 7
  □ □ , □ □ □ , □ □ □ , □ □ □
```

Now, try setting up the addition on your own.

We usually write the number with more digits on top.

This makes it easier to line up the digits by place value.

PRACTICE | Compute each sum below.

27. $1,659 + 282 =$ _____

28. $746 + 1,297 =$ _____

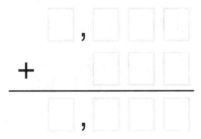

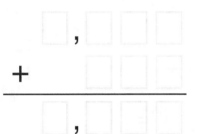

29. $531 + 284 =$ _____

30. $538 + 765 =$ _____

31. $8,376 + 247 =$ _____

32. $527 + 1,934 =$ _____

EXAMPLE | Fill each empty box with a digit to give a correct sum.

$$\begin{array}{r} \square\,7 \\ +\ 4\,6 \\ \hline 1\,2\,\square \end{array}$$

In the ones column, since 7+6 = 1$\boxed{3}$, the ones digit of the sum is **3**, and we place a 1 above the tens column.

$$\begin{array}{r} {}^{1}\ \ \\ \square\,7 \\ +\ 4\,6 \\ \hline 1\,2\,\boxed{3} \end{array}$$

In the tens column, since 1+$\boxed{7}$+4 = 12, we place **7** in the remaining empty box.

$$\begin{array}{r} {}^{1}\ \ \\ \boxed{7}\,7 \\ +\ 4\,6 \\ \hline 1\,2\,\boxed{3} \end{array}$$

PRACTICE | Fill each empty box below with a digit to give a correct sum.

33.
$$\begin{array}{r} \square\ 7\ 2 \\ +\ 3\ \square\ 5 \\ \hline 9\ 8\ \square \end{array}$$

34.
$$\begin{array}{r} 5\ \square\ 6 \\ +\ \square\ 4\ 9 \\ \hline 7\ 8\ \square \end{array}$$

35.
$$\begin{array}{r} 7\ 0\ \square \\ +\ \square\ 3\ 4 \\ \hline 1{,}5\ \square\ 7 \end{array}$$

36.
$$\begin{array}{r} 9\ 2\ \square \\ +\ \square\ 2\ 4 \\ \hline \square{,}5\ \square\ 9 \end{array}$$

37.
$$\begin{array}{r} \square\ 4\ 6 \\ +\ 5\ \square\ 4 \\ \hline 8\ 4\ \square \end{array}$$

38.
$$\begin{array}{r} 7\ 0\ \square \\ +\ \square\ \square\ 8 \\ \hline 1{,}0\ 5\ 3 \end{array}$$

PRACTICE | Fill each empty box below with a digit to give a correct sum.

39.
```
    7 , □ 0 □
  +   □ , 8 6 4
  ─────────────
    1 4 , 2 □ 9
```

40.
```
    8 , 6 4 □
  +   □ , □ 8
  ─────────────
    1 1 , 2 6 3
```

41.
```
    7 , □ □ □
  +   □ , 8 8 8
  ─────────────
    1 0 , 3 3 3
```

42.
```
    4 , 2 □ □
  +   □ , □ 1 9
  ─────────────
    6 , 1 3 3
```

43.
```
    1 , □ 3 □
  +   □ , 8 1 8
  ─────────────
    1 □ , 8 □ 4
```

44.
```
    □ , □ 2 8
  +   5 , 4 □ □
  ─────────────
    □ 1 , 1 7 6
```

45. ★
```
    □ , □ □ □
  +       6 5 4
  ─────────────
    □ □ , 5 4 3
```

46. ★
```
    7 , □ 0 □
  +   □ , 8 0 1
  ─────────────
    1 7 , 1 1 □
```

The algorithm works for adding more than two numbers, too.

EXAMPLE | Compute 148+382+275.

We stack the addition so that the place values line up. Then, we add each place value the same way we do when adding two numbers.

Add the ones.	Add the tens.	Add the hundreds.
$\overset{1}{}$	$\overset{2}{}\overset{1}{}$	$\overset{2}{}\overset{1}{}$
1 4 8	1 4 8	1 4 8
3 8 2	3 8 2	3 8 2
+ 2 7 5	+ 2 7 5	+ 2 7 5
5	0 5	8 0 5

So, 148+382+275 = **805**.

PRACTICE | Compute each sum below.

47.
```
    8 4
    3 3
  + 7 1
  ───────
  □ □ □
```

48.
```
    3 4 9
    2 5 8
  +   6 7
  ───────
  □ □ □
```

49.
```
    8 4 1
    3 5 8
  + 9 3 4
  ───────
  □ , □ □ □
```

50.
```
    6 5 1
    5 2 5
  +   7 4
  ───────
  □ , □ □ □
```

51.
```
    3 8 1
    5 6 9
  + 7 5 2
  ───────
  □ , □ □ □
```

52.
```
    4 5 6
    7 7 7
  + 1 4 9
  ───────
  □ , □ □ □
```

PRACTICE | Compute each sum below.

53. 59 + 82 + 16 + 37 = _____

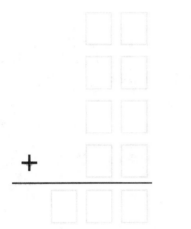

54. 742 + 5,389 + 67 = _____

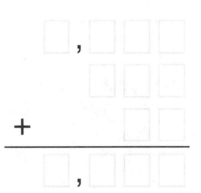

55. 531 + 284 + 672 = _____

56. 807 + 1,049 + 425 = _____

Need more practice? Print more problems at BeastAcademy.com!

EXAMPLE | Fill the blanks using the digits 2, 3, 5, and 7 once each to give a correct sum.

```
    □ 3
    9 6
  +  6 □
  ─────
   □ □ 4
```

In the ones column, since 3+6+$\boxed{5}$ = 14, the missing ones digit is **5**. We also place 1 more ten in the tens column.

```
      1
    □ 3
    9 6
  +  6 5
  ─────
   2 □ 4
```

The largest sum we can get in the tens column is 1+7+9+6 = 23. So, the hundreds digit of the sum is 1 or 2. Only 2 is in the list of missing digits, so the hundreds digit of the sum is **2**.

The remaining digits are 3 and 7.

In the tens column, only 1+$\boxed{7}$+9+6 = 2$\boxed{3}$ gives a true statement, so we place the **7** and the **3** as shown.

```
      1
    7 3
    9 6
  +  6 5
  ─────
   2 3 4
```

Check: 73+96+65 = 234. ✓

PRACTICE | For each problem below, use the given digits once each to fill the blanks and give a correct sum.

57. **Missing digits:** 1, 5, 6

```
     □
  +  9
  ────
   □ □
```

58. **Missing digits:** 0, 4, 6

```
   1 □ 4
  +  5 □
  ─────
   2 0 □
```

PRACTICE | For each problem below, use the given digits once each to fill the blanks and give a correct sum.

59. **Missing digits:** 0, 1, 8, 9

$$
\begin{array}{r}
3\ 4\ \square \\
+\ 5\ 5\ 3 \\
\hline
\square\ \square\ \square
\end{array}
$$

60. **Missing digits:** 4, 5, 8, 9

$$
\begin{array}{r}
1\ 4\ \square \\
+\ 3\ \square\ 7 \\
\hline
\square\ \square\ 5
\end{array}
$$

61. **Missing digits:** 1, 2, 3, 8, 9

$$
\begin{array}{r}
8\ 1\ \square \\
+\ \square\ \square\ 5 \\
\hline
\square\ 4\ \square
\end{array}
$$

62. ★ **Missing digits:** 0, 2, 3, 8, 9

$$
\begin{array}{r}
\square\,6\ 2\ \square \\
+\ 2\,7\ \square\ 1 \\
\hline
1\ \square\,4\ \square\ 4
\end{array}
$$

63. ★ **Missing digits:** 0, 1, 2, 6, 7

$$
\begin{array}{r}
5\ \square\ 6 \\
1\ 5\ \square \\
+\ \square\ 4\ 6 \\
\hline
\square\,4\ 6\ \square
\end{array}
$$

64. ★ **Missing digits:** 0, 2, 4, 5, 8

$$
\begin{array}{r}
\square\,0\ 2\ 9 \\
6\,1\ 9\ \square \\
+\ 1\,1\ \square\ 1 \\
\hline
1\ \square\,4\ \square\ 4
\end{array}
$$

In a **Cross-Sums** puzzle, the goal is to fill each circle with a different digit to create two sums. One sum can be read by tilting the grid to the left, and the other can be read by tilting the grid to the right, as shown in the example below.

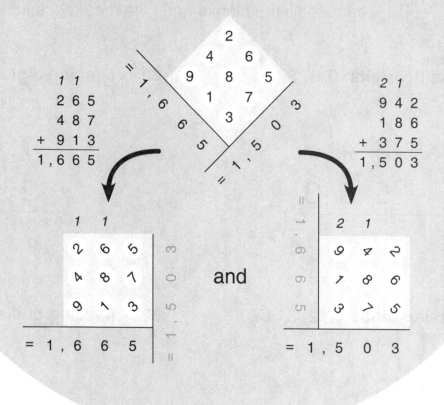

PRACTICE | Complete each Cross-Sums below so that the digits *1 through 4* are each used once in the circles.

65.

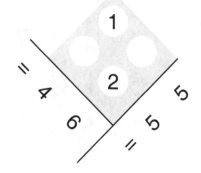

66.

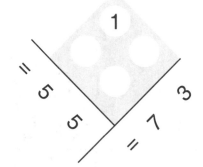

PRACTICE Complete each Cross-Sums below so that the digits *1 through 9* are each used once in the circles.

67.

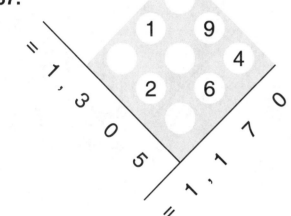

68.

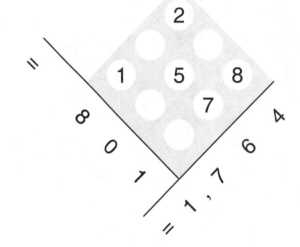

69.

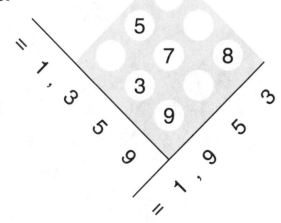

70.

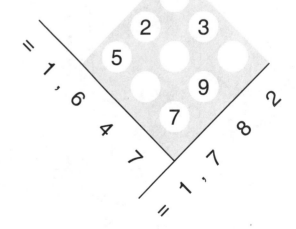

PRACTICE | Complete each Cross-Sums below so that the digits *1 through 9* are each used once in the circles.

71.

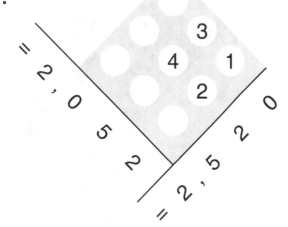

72.

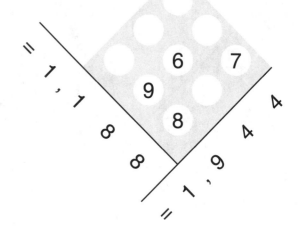

73.

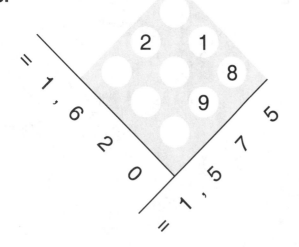

74.

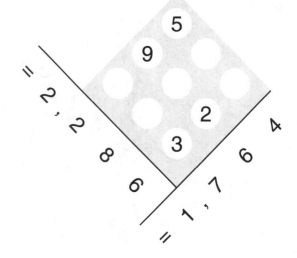

PRACTICE Complete each Cross-Sums below so that the digits *1 through 9* are each used once in the circles.

75.
★

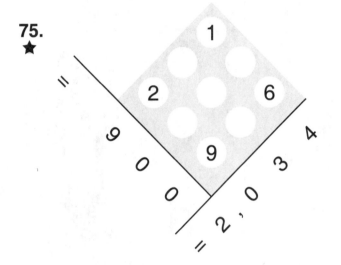

76.
★

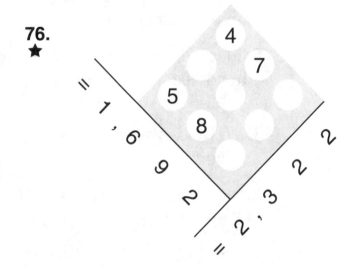

77.
★

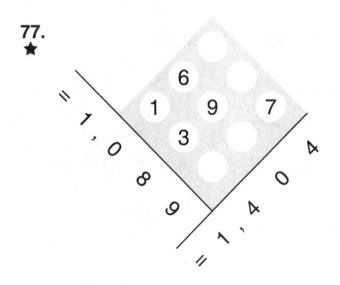

78.
★
★

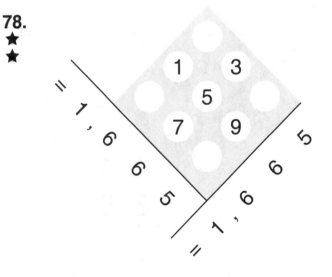

EXAMPLE | Compute 556−274.

We stack the numbers and subtract by place value from **right** to **left**. In the ones column, we have 6−4 = 2.

$$\begin{array}{r} 5\ 5\ 6 \\ -\ 2\ 7\ 4 \\ \hline 2 \end{array}$$

In the tens column, we can't take 7 away from 5. So, we break a hundred to make 10 more tens.

5 hundreds and 5 tens is the same as 4 hundreds and 15 tens.

$$\begin{array}{r} {}^{4}\ {}^{15} \\ \cancel{5}\ \cancel{5}\ 6 \\ -\ 2\ 7\ 4 \\ \hline 2 \end{array}$$

Then we can subtract the tens and the hundreds.

In the tens column, we have 15−7 = 8. In the hundreds column, we have 4−2 = 2.

So, 556−274 = **282**.

$$\begin{array}{r} {}^{4}\ {}^{15} \\ \cancel{5}\ \cancel{5}\ 6 \\ -\ 2\ 7\ 4 \\ \hline 2\ 8\ 2 \end{array}$$

We can also stack subtraction.

PRACTICE | Fill the empty boxes to compute each difference below.

79. 73−36

$$\begin{array}{r} 7\ 3 \\ -\ 3\ 6 \\ \hline \end{array}$$

80. 547−65

$$\begin{array}{r} 5\ 4\ 7 \\ -\ \ \ 6\ 5 \\ \hline \end{array}$$

PRACTICE | Compute each difference below, breaking as needed.

81.
$$
\begin{array}{r}
3\ 8\ 2 \\
-\ \ 5\ 4 \\
\hline
\square\ \square\ \square
\end{array}
$$

82.
$$
\begin{array}{r}
9\ 2\ 7 \\
-\ 1\ 5\ 5 \\
\hline
\square\ \square\ \square
\end{array}
$$

83.
$$
\begin{array}{r}
7\ 5\ 8 \\
-\ 4\ 3\ 9 \\
\hline
\square\ \square\ \square
\end{array}
$$

84.
$$
\begin{array}{r}
8\ 0\ 7 \\
-\ 3\ 6\ 4 \\
\hline
\square\ \square\ \square
\end{array}
$$

85.
$$
\begin{array}{r}
5,6\ 4\ 3 \\
-\ 1,9\ 2\ 1 \\
\hline
\square,\square\ \square\ \square
\end{array}
$$

86.
$$
\begin{array}{r}
9,3\ 8\ 6 \\
-\ 7,1\ 9\ 6 \\
\hline
\square,\square\ \square\ \square
\end{array}
$$

Subtraction

EXAMPLE | Compute 907−348.

In the ones place, we can't take 8 away from 7.
So, we need to break a ten to make 10 more ones.
But, there are 0 tens in 907.

So, we first break a hundred to make 10 tens.

9 hundreds and 0 tens is the same as
8 hundreds and 10 tens.

$$\begin{array}{r} {}^{8}\;{}^{10} \\ \cancel{9}\;\cancel{0}\;7 \\ -\;3\;4\;8 \\ \hline \end{array}$$

Now, we can break a ten to make 10 ones.

10 tens and 7 ones is the same as
9 tens and 17 ones.

$$\begin{array}{r} {}^{9} \\ {}^{8}\;\cancel{\cancel{0}}\;17 \\ \cancel{9}\;\cancel{0}\;\cancel{7} \\ -\;3\;4\;8 \\ \hline \end{array}$$

Finally, we subtract by place value.
In the ones column, 17−8=9.
In the tens column, 9−4=5.
In the hundreds column, 8−3=5.
So, 907−348=**559**.

$$\begin{array}{r} {}^{9} \\ {}^{8}\;\cancel{\cancel{0}}\;17 \\ \cancel{9}\;\cancel{0}\;\cancel{7} \\ -\;3\;4\;8 \\ \hline 5\;5\;9 \end{array}$$

> Sometimes we need to break more than one place value.

PRACTICE | Compute each difference below.

87.
$$\begin{array}{r} 8\;5\;7 \\ -\;3\;6\;9 \\ \hline \end{array}$$

88.
$$\begin{array}{r} 5\;1\;4 \\ -\;3\;8\;6 \\ \hline \end{array}$$

89.
$$\begin{array}{r} 2\;0\;3 \\ -\;\;\;5\;9 \\ \hline \end{array}$$

PRACTICE | Compute each difference below.

90. $443 - 67 =$ _____

91. $5,806 - 567 =$ _____

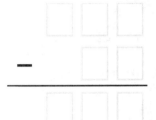

92. $916 - 248 =$ _____

93. $2,736 - 829 =$ _____

94. $3,067 - 2,278 =$ _____

95. $40,061 - 13,456 =$ _____

Need more practice? Print more problems at BeastAcademy.com!

You can stack addition or subtraction problems...

...but stacking isn't always a great way to add or subtract.

PRACTICE | Solve each problem below.

96.
$$\begin{array}{r} 5\ 2\ 1 \\ -\quad 9\ 4 \\ \hline \end{array}$$

97. $521 - 94 = 521 - 100 + \underline{}$

$= \underline{}$

98.
$$\begin{array}{r} 3\ 9\ 6 \\ +\quad 8\ 7 \\ \hline \end{array}$$

99. $396 + 87 = 400 + \underline{}$

$= \underline{}$

100.
$$\begin{array}{r} 8\ 5\ 7 \\ -\quad 5\ 9 \\ \hline \end{array}$$

101. $857 - 59 = 857 - 57 - \underline{}$

$= \underline{}$

PRACTICE | Solve each problem below.

102.
$$\begin{array}{r} 1\ 9\ 7 \\ 1\ 9\ 8 \\ +\ 1\ 9\ 9 \\ \hline \end{array}$$

103. $197+198+199 = 600 - \underline{\hspace{1cm}}$

$= \underline{\hspace{1cm}}$

104.
$$\begin{array}{r} 2,6\ 7\ 5 \\ +\ \ \ \ 9\ 9\ 8 \\ \hline \end{array}$$

105. $2,675+998 = 2,675+1,000 - \underline{\hspace{1cm}}$

$= \underline{\hspace{1cm}}$

106.
$$\begin{array}{r} 1,0\ 0\ 0 \\ -\ \ \ \ 8\ 8\ 7 \\ \hline \end{array}$$

107. $887+\underline{\hspace{1cm}} = 1,000.$

So, $1,000-887 = \underline{\hspace{1cm}}.$

108.
$$\begin{array}{r} 8,0\ 0\ 2 \\ -\ 7,9\ 9\ 6 \\ \hline \end{array}$$

109. $7,996+\underline{\hspace{1cm}} = 8,002.$

So, $8,002-7,996 = \underline{\hspace{1cm}}.$

Try these **without** stacking.

PRACTICE | Compute each sum or difference below.

110. $100{,}013 - 99{,}996 =$ _____

111. $1{,}980 + 2{,}220 =$ _____

112. $8{,}817 - 819 =$ _____

113. $4{,}000 - 1{,}725 =$ _____

114. $9{,}990 + 9{,}995 =$ _____

115. $2{,}772 - 596 =$ _____

116. $3{,}333 + 2{,}980 =$ _____

117. $3{,}975 + 625 =$ _____

118. $53{,}211 - 53{,}209 =$ _____

119. $9{,}999 + 9{,}998 + 9{,}997 =$ _____

Addition and subtraction are related.
If $10-3=7$, then $7+3=10$. So, we can check a subtraction problem using addition.

For example, we can check $815-578=237$ by adding $237+578$ to see if we get 815.

$$
\begin{array}{r}
{\scriptstyle 10} \\
7\ \cancel{8}\ 15 \\
\cancel{8}\ \cancel{1}\ \cancel{5} \\
-\ 5\ 7\ 8 \\
\hline
2\ 3\ 7
\end{array}
\qquad\Rightarrow\qquad
\begin{array}{r}
{\scriptstyle 1\ 1} \\
2\ 3\ 7 \\
+\ 5\ 7\ 8 \\
\hline
8\ 1\ 5
\end{array}
$$

Since $237+578=815$, $815-578=237$ is correct! ✓

PRACTICE | Use addition to check the subtraction below.

120. Cross out any subtraction below that is incorrect.

$$
\begin{array}{r}
5\ 2\ 9 \\
-\ 1\ 4\ 6 \\
\hline
3\ 8\ 3
\end{array}
\qquad
\begin{array}{r}
7\ 1\ 6 \\
-\ 3\ 4\ 4 \\
\hline
3\ 6\ 2
\end{array}
\qquad
\begin{array}{r}
5\ 1\ 8 \\
-\ 3\ 4\ 3 \\
\hline
1\ 7\ 5
\end{array}
$$

121. Cross out any subtraction below that is incorrect.

$$
\begin{array}{r}
7\ 0\ 6 \\
-\ 2\ 3\ 8 \\
\hline
4\ 5\ 8
\end{array}
\qquad
\begin{array}{r}
8\ 0\ 4 \\
-\ 3\ 2\ 7 \\
\hline
4\ 7\ 7
\end{array}
\qquad
\begin{array}{r}
6\ 0\ 5 \\
-\ 3\ 4\ 6 \\
\hline
3\ 6\ 9
\end{array}
$$

122. Cross out any subtraction below that is incorrect.

$$
\begin{array}{r}
5\ 1\ 3 \\
-\ 3\ 5\ 7 \\
\hline
1\ 5\ 6
\end{array}
\qquad
\begin{array}{r}
7\ 5\ 3 \\
-\ 1\ 7\ 5 \\
\hline
5\ 7\ 2
\end{array}
\qquad
\begin{array}{r}
9\ 3\ 1 \\
-\ 5\ 7\ 7 \\
\hline
3\ 5\ 4
\end{array}
$$

PRACTICE | Answer each question below.

123. Mount Cyclops is 3,427 meters tall.
Mount Scylla is 4,184 meters tall.
How many meters taller is Mount Scylla
than Mount Cyclops?

123. _____

124. Ernie's Omelet Shop ordered 1,344 eggs last week.
Unfortunately, a crate containing 576 eggs fell, breaking
all of the eggs inside. How many of the 1,344 eggs were
unbroken?

124. _____

125. 2,589 years ago, the Behemoth Pine tree was 1,478
years old. How many years old is this tree now?

125. _____

126. There are 6,435 ways to choose 7 of the 15 available
toppings for a pizza at Hutt's Pizza. There are 1,365
possible 4-topping pizzas. How many more 7-topping
pizzas are possible than 4-topping pizzas?

126. _____

PRACTICE | Answer each question below.

127. A box of Beastie Bites cereal contains 567 grams of cereal. How many grams of cereal are in 3 boxes of Beastie Bites?

127. _____

128. The trip from Hydra Harbor to Cape Capricorn is 2,250 miles by boat and 2,827 miles by car. If you travel from Hydra Harbor to Cape Capricorn *and back*, how many miles shorter is the trip by boat than by car?

128. _____

129. Maude the Marauder's pirate ship weighs 1,456 tons. Cutlass Pete's ship weighs 300 tons *more* than Maude's. How many tons do these two ships weigh together?

129. _____

130. ★ Borg has exactly 3,000 Beastball cards in his collection. 2,156 of these cards are in a shoebox, and the rest are in a carrying case. How many *more* Beastball cards are in Borg's shoebox than in his carrying case?

130. _____

In a **Cryptarithm** puzzle, each letter stands for a digit.

- The same letter always stands for the same digit. For example, if an A stands for 1, then all A's in that puzzle stand for 1.

- Two different letters stand for different digits. For example, A and B cannot both stand for 1.

The goal is to figure out which digit each letter stands for.

EXAMPLE	Find the digit that each letter stands for in the Cryptarithm puzzle to the right.

$$\begin{array}{r} A\ A \\ +\ A\ A \\ \hline B\ B\ C \end{array}$$

The largest sum of two 2-digit numbers is $99+99=198$. So, the hundreds digit of the sum must be 1.

That means B is 1. We replace both B's with 1's.

$$\begin{array}{r} A\ A \\ +\ A\ A \\ \hline 1\ 1\ C \end{array}$$

In the tens column, we cannot add a number to itself and get 11. But, if the sum of the ones column gives an extra ten, we can get 11 in the tens column by adding $1+5+5$. That means A is 5. We replace the A's with 5's.

$$\begin{array}{r} 1\ \ \\ 5\ 5 \\ +\ 5\ 5 \\ \hline 1\ 1\ C \end{array}$$

Finally, in the ones column, since $5+5=10$, C is 0.

So, A = **5**, B = **1**, and C = **0**.

We check our work: $55+55=110$. ✓

$$\begin{array}{r} 1\ \ \\ 5\ 5 \\ +\ 5\ 5 \\ \hline 1\ 1\ 0 \end{array}$$

PRACTICE	Solve each Cryptarithm puzzle below.

131.

$$\begin{array}{r} \boxed{\ }\ 2 \\ +\ \boxed{\ }\ 2 \\ \hline \boxed{\ }\ \boxed{\ } \end{array}$$

A = ____

B = ____

132.

$$\begin{array}{r} \boxed{\ }\ 5 \\ +\ 4\ \boxed{\ } \\ \hline \boxed{\ }\ 8 \end{array}$$

A = ____

B = ____

PRACTICE | Solve each Cryptarithm puzzle below.

133.
```
    B A 5
  + C 6 5
  ─────────
    9 B A
```
A = ___
B = ___
C = ___

134.
```
    B A A
  +   4 2
  ─────────
  A , C B 3
```
A = ___
B = ___
C = ___

135.
```
    C A 8
  + B A 8
  ─────────
    C B A
```
A = ___
B = ___
C = ___

136.
```
    A B 3
  +   A 5 5
  ─────────
  C , B C B
```
A = ___
B = ___
C = ___

137. ★
```
    A A A
  +   B A B
  ─────────
    C B A
```
A = ___
B = ___
C = ___

138. ★
```
    B A C
  +   B C B
  ─────────
  C , A C C
```
A = ___
B = ___
C = ___
D = ___

Print these and more Cryptarithms at BeastAcademy.com!

ALGORITHMS

PRACTICE | Answer each question below.

139. What is the smallest number that can be
added to 8,325 to give a 5-digit result?

139. _____

140. Fill the blank below to make a true equation.
★

$$123,456 - \text{_____} = 45,678$$

141. Find the sum $1+11+111+1,111+11,111+111,111$.
Try to answer without stacking the addition.

141. _____

142. $1,234 + 2,345 + 3,456 + 4,567 + 5,678 = 17,280$.
★   What is $2,345 + 3,456 + 4,567 + 5,678 + 6,789$?

142. _____

PRACTICE | Answer each question below.

143. ★ Fill every blank below with the *same* digit to make a true statement.

144. ★ Grogg adds two numbers that only use the digit 5. The thousands digit of the sum is 6. What is the smaller of the two numbers Grogg adds?

144. _____

145. ★★ Lizzie uses the digits 0 through 7 once each to make two 4-digit numbers. What is the smallest possible difference between these two numbers?

145. _____

146. ★★ What digits do A, B, and C stand for if A,AAA + B,BBB + C,CCC = AB,BBC?

146. A = ___

B = ___

C = ___

CHAPTER 12
Problem Solving

This is a very challenging chapter that will help you learn some great ways to solve tough problems.

Take your time, and don't worry if you can't solve many of the problems on your first try!

Use this Practice book with Guide 2D from BeastAcademy.com.

Recommended Sequence:

Book	Pages:
Guide:	76-81
Practice:	71-85
Guide:	82-93
Practice:	86-91
Guide:	94-107
Practice:	92-105

You may also read the entire chapter in the Guide before beginning the Practice chapter.

Organization can make many problems easier to solve.

Are some of the problems below easier than others?

PRACTICE | Solve each problem below.

1. Circle the third-largest number in the list below.

777   778   787   788   877   878   887   888

2. Circle the third-largest number in the list below.

647   644   690
719   689   646
588   542

3. The list below of numbers from 30 to 42 is missing two numbers. Which two numbers are missing from the list?

3. _____ & _____

30   31   33   34   35   36   37   39   40   41   42

4. The list below of numbers from 45 to 57 is missing two numbers. Which two numbers are missing from the list?

4. _____ & _____

50   53   47   46   51   57   56   55   52   48   45

EXAMPLE | Count the number of triangles in the diagram below.

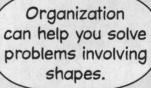

Organization can help you solve problems involving shapes.

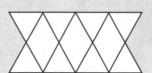

There are lots of triangles in the diagram. We can organize our work as shown below to make sure we count each triangle exactly once.

First, there are small triangles. Four triangles point up, and four triangles point down. So, there are a total of 8 small triangles.

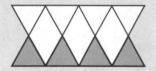 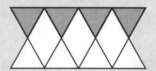

Then, there are large triangles. These triangles overlap and are harder to count. Three large triangles point up, and three point down. So, there are a total of 6 large triangles.

This gives a total of 8+6 = **14** triangles.

PRACTICE | Count the number of squares in the diagram below.

5.   Small: _____

          Large: _____

          Total: _____

PRACTICE | Count the number of each shape in the diagrams below.

6. Triangles

Small: _____

Large: _____

Total: _____

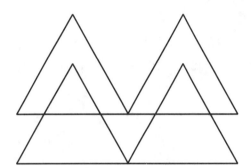

7. Triangles

Small: _____

Large: _____

Total: _____

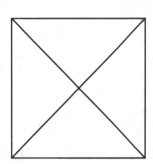

8. Squares: _____

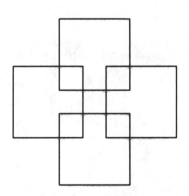

9. ★ Triangles: _____

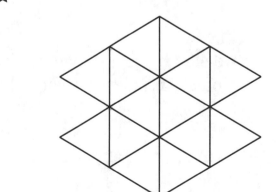

A **Taxi Path** connects two dots on a city grid. Starting at the dot in the top-left, the path can only go *down* or *right* to reach the dot in the bottom-right. The goal of a Taxi Path puzzle is to count the total number of possible paths on a given grid.

EXAMPLE | How many different Taxi Paths connect the two dots on the grid shown?

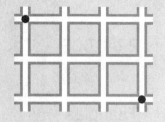

To make sure we find all of the paths, we organize our work as shown below. If we start by going three blocks to the right, then there is only one way to finish.

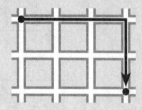

If we start by going exactly two blocks to the right, we have to turn down next. Then, there are two ways to complete the path.

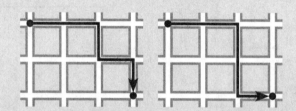

If we start by going exactly one block to the right, we have to turn down next. Then, there are three ways to complete the path.

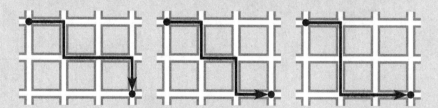

If we start by going down instead of right, then there are four ways to complete the path. This gives a total of 1+2+3+4 = **10** paths.

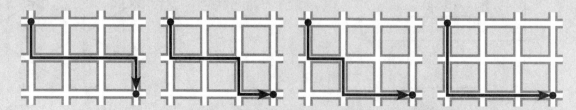

PRACTICE Draw all of the paths on each Taxi Path puzzle below.

10. There are three Taxi Paths that connect the top-left dot to the bottom-right dot. Draw all three paths on the grids below.

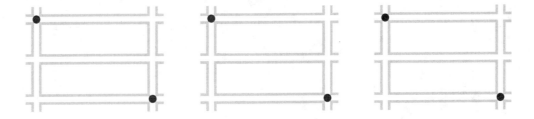

11. There are four Taxi Paths that connect the top-left dot to the bottom-right dot. Draw all four paths on the grids below.

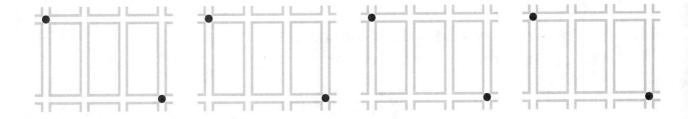

12. There are four Taxi Paths that connect the top-left dot to the bottom-right dot. Draw all four paths on the grids below.

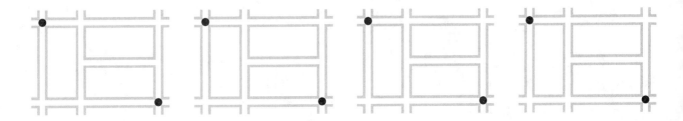

PRACTICE | Solve each Taxi Path puzzle below.

13. How many different Taxi Paths can be drawn on the grids below? There may be less than eight.

13. _____

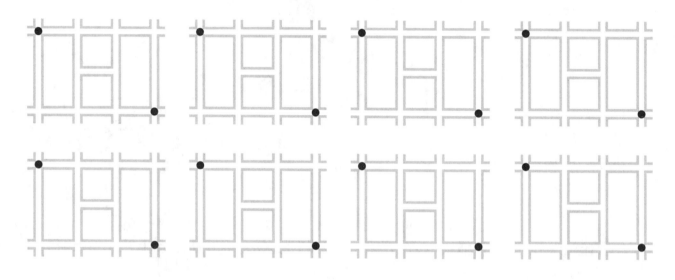

14. How many different Taxi Paths can be drawn on the grids below? There may be less than eight.

14. _____

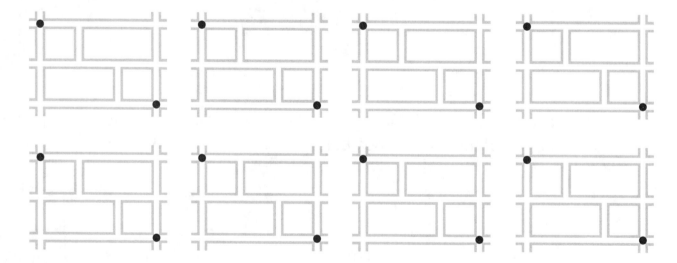

PRACTICE | Solve each Taxi Path puzzle below.

15. How many different Taxi Paths can be drawn on the grids below? There may be less than eight.

15. _____

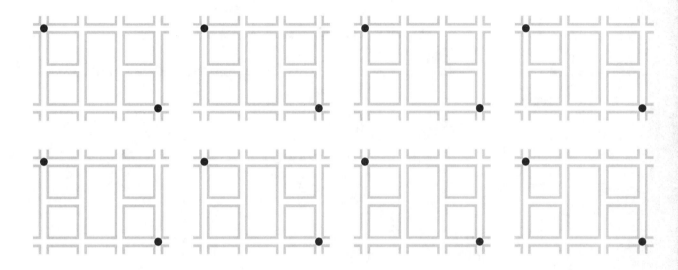

16. ★ How many different Taxi Paths can be drawn on the grids below? There may be less than eight.

16. _____

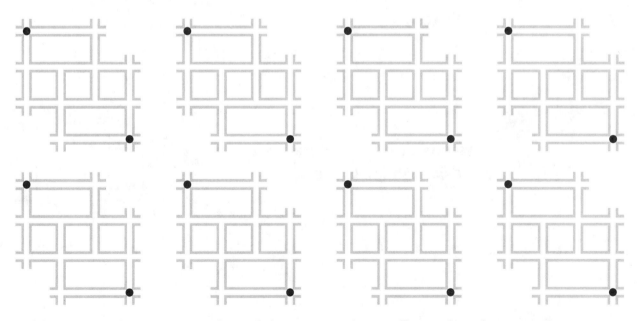

You can print more copies of these puzzles at BeastAcademy.com.

A **Checkerboard Path** passes through every square on a checkerboard exactly once. Paths can only go up, down, left, or right (not diagonally).

EXAMPLE | How many different Checkerboard Paths begin in the square marked below and visit every square exactly once?

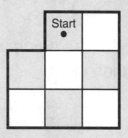

We check every option as shown below, being careful not to miss any.

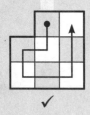
✓

> We start by going either down or right.

> If we go down first, then there is only one way to finish the path.

> If we go right first, we must go down, then either left or down.

> If we go left, then there is only one way to finish.

> If we go down, then there are two ways we can finish the path.

✓

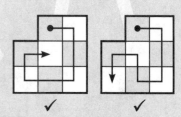
✓   ✓

So, there are a total of **4** Checkerboard Paths, all marked with ✓'s above.

PRACTICE | Solve each Checkerboard Path problem below.

17. There are two ways to complete a Checkerboard Path on the board below, starting at the dot. Draw both paths.

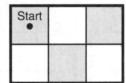 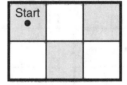

18. There are three ways to complete a Checkerboard Path on the board below, starting at the dot. Draw all three paths.

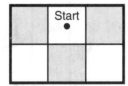 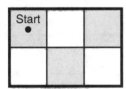 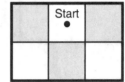

19. There are three ways to complete a Checkerboard Path on the board below, starting at the dot. Draw all three paths.

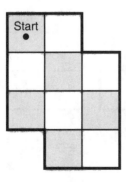 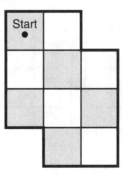 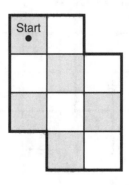

PRACTICE | Solve each Checkerboard Path problem below.

20. How many different Checkerboard Paths can be drawn on the grid below? There may be less than five.

20. _____

21. How many different Checkerboard Paths can be drawn on the grid below? There may be less than five.

21. _____

22. How many different Checkerboard Paths can be drawn on the grid below? There may be less than five.

22. _____

PRACTICE | Solve each Checkerboard Path problem below.

23. How many different Checkerboard Paths can be drawn on the grid below? There may be less than eight.

23. _____

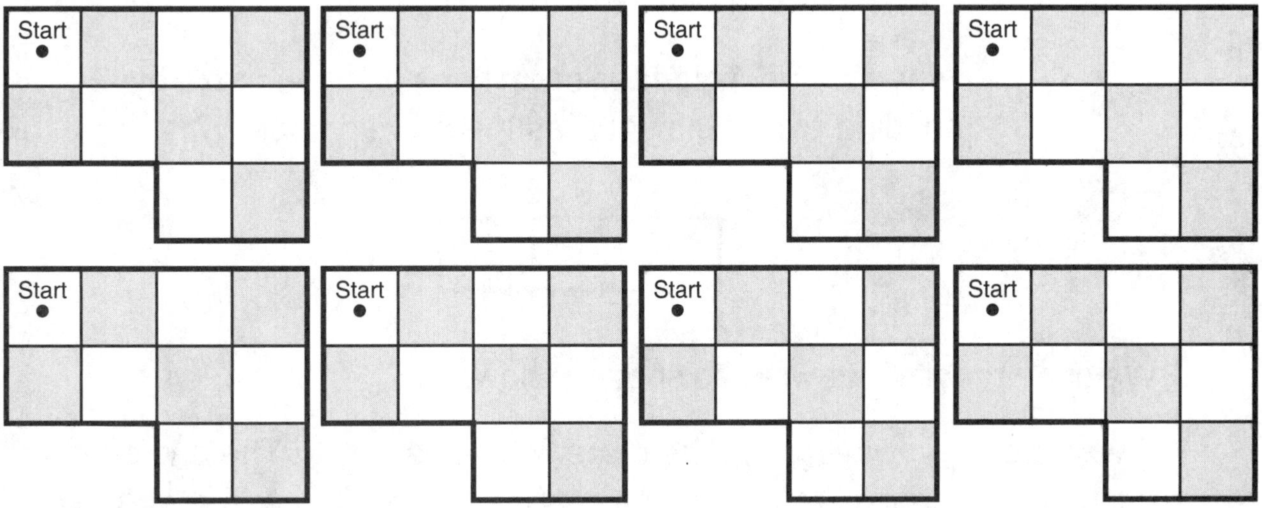

24. How many different Checkerboard Paths can be drawn on the grid below? There may be less than eight.

24. _____

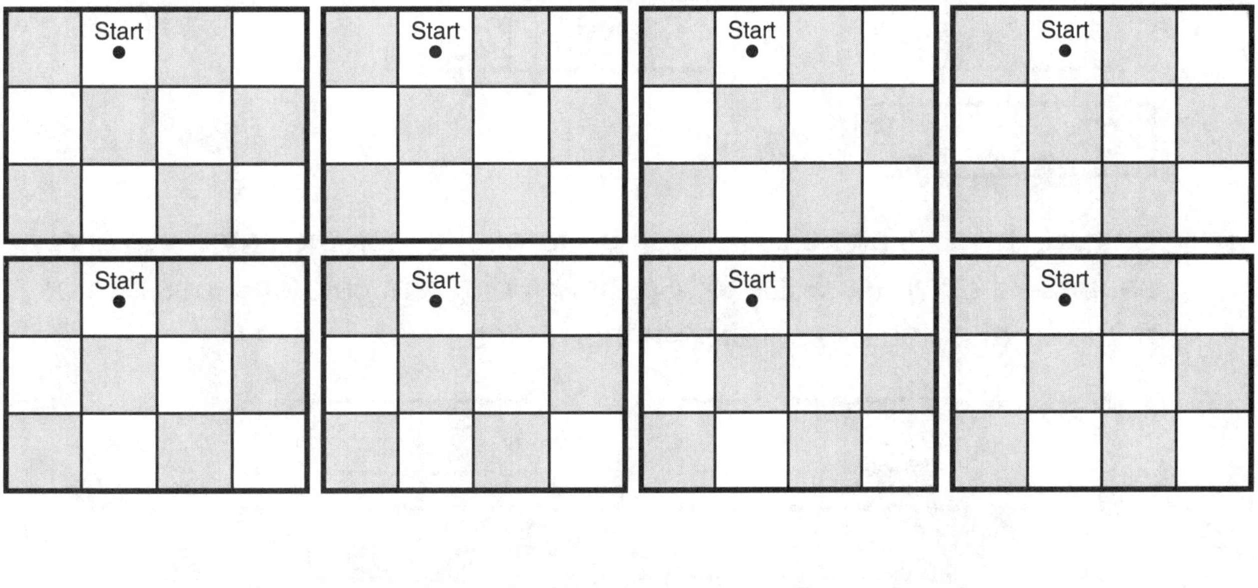

You can print more copies of these puzzles at BeastAcademy.com.

A **coloromino** is a shape made of shaded and unshaded squares.

Two shapes are the same coloromino if one can be turned to look like the other.

These two colorominoes are the same, since you can turn one to look like the other.

These two colorominoes are **not** the same, since you cannot turn one to look like the other.

✓

✗

EXAMPLE How many **different** colorominoes can you make by shading **two** of the squares in the shape below?

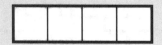

We can organize our work as shown below.

| There are three ways to shade the shape that have the far-left square shaded. | There are two ways that have the second square shaded, but not the first. | There is only one way that does not shade either of the first two squares. |

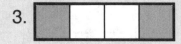

Shapes 1 (⬚⬚▢▢) and 6 (▢▢⬚⬚) can be turned to look the same. So can shapes 2 (⬚▢⬚▢) and 5 (▢⬚▢⬚). So, only **4** different colorominoes can be made by shading two squares in this shape.

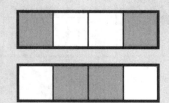

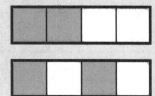

PRACTICE | Solve each coloromino problem below.

25. Show all three ways to shade **two** squares in the shape below.

26. How many **different** colorominoes did you make?

26. _____

27. Show all four ways to shade **one** square on the shape below.

28. How many **different** colorominoes did you make?

28. _____

29. Show all six ways to shade **two** squares on the shape below.

30. How many **different** colorominoes did you make?

30. _____

PRACTICE | Solve each Coloromino problem below.

31. Show all six ways to shade *five* squares on the shape below.

32. How many *different* colorominoes did you make? 32. _____

33. Show all six ways to shade *two* squares on the shape below.

34. How many *different* colorominoes did you make? 34. _____

35. Show all six ways to shade *two* squares on the shape below.

36. How many *different* colorominoes did you make? 36. _____

PRACTICE | Solve each Coloromino problem below.

37. ★ Show all ten ways to shade *two* squares on the shape below.

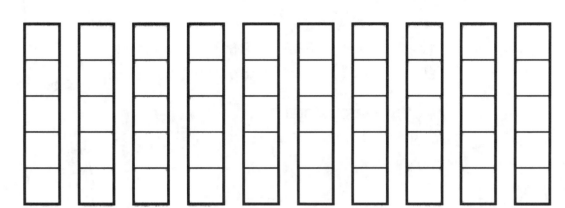

38. ★ How many *different* colorominoes did you make?

38. _____

39. ★ There are ten ways to shade *three* squares on the shape below. There are only three different colorominoes that can be made. Shade three squares in each of the shapes below to make three *different* colorominoes.

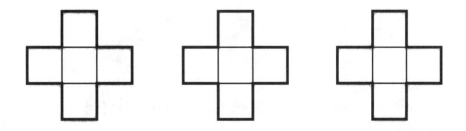

EXAMPLE | How many different two-digit numbers can be made using only the digits 2, 4, 6, and 8?

We can write the numbers in order.

We organize the numbers in the 20's, 40's, 60's, and 80's as shown below.

22	42	62	82
24	44	64	84
26	46	66	86
28	48	68	88

So, there are a total of
4+4+4+4 = **16** numbers.

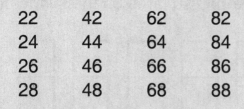

If you make a list to solve a problem, keep the list organized.

PRACTICE | Solve each problem below.

40. There are four 2-digit numbers that use only 6's and 7's.
List them in order from least to greatest below.

_____, _____, _____, _____

41. There are eight numbers less than 1,000 that use only 1's and 0's as digits.
Complete the list of these numbers in order below.

___0___, _____, _____, _____, _____, _____, _____, __111__

42. How many numbers between 200 and 400 use only 2's and 3's as digits?

42. _____

In the problems on this page, you are **arranging** digits.

PRACTICE | Solve each problem below.

43. There are four ways to arrange the digits 5, 5, 5, and 7 to make a four-digit number. List them in order from least to greatest below.

_____ , _____ , _____ , _____

44. There are six ways to arrange the digits 2, 3, and 4 to make a three-digit number. List them in order from least to greatest below.

_____ , _____ , _____ , _____ , _____ , _____

45. ★ There are six ways to arrange the digits 0, 0, 1 and 2 to make a four-digit number. List them in order from least to greatest below.

_____ , _____ , _____ , _____ , _____ , _____

46. ★ How many different four-digit numbers can you make by arranging the digits 1, 1, 9, and 9?

46. _____

EXAMPLE

How many ways can we add 1's, 2's, and 3's to get a sum of 5 if we don't care about the order of the numbers? For example, 2+3 and 3+2 are counted as the same.

We can organize our work by the largest number in the sum.

If 3 is the largest number in the sum, there are two ways to get a sum of 5.

$$3+2$$
$$3+1+1$$

If 2 is the largest number, there are two ways.

$$2+2+1$$
$$2+1+1+1$$

If 1 is the largest number, there is just one way.

$$1+1+1+1+1$$

All together, there are **5** ways.

PRACTICE

Solve the problem below. In this section, we don't care about the order of the numbers we're adding. For example, 4+5 and 5+4 are counted as the same.

47. There are four ways to get a sum of 7 using 1's and 2's. Write all four ways on the lines below.

_____

_____

_____

_____

PRACTICE | Solve each problem below. For the problems in this section, we don't care about the order of the numbers. For example, 4+5 and 5+4 are counted as the same.

48. There are four ways to get a sum of 8 using 2's, 3's, and 4's. Write all four ways on the lines below.

_____

_____

_____

_____

49. How many ways are there to get a sum of 10 using 1's and 3's?

49. _____

50. How many ways are there to get a sum of 8 using 1's, 3's, and 5's?

50. _____

51. ★ How many ways are there to get a sum of 6 using 1's, 2's, and 3's?

51. _____

Spotting patterns can help you solve lots of problems.

Here are some types of patterns you should be able to recognize.

Below are some common patterns to look for.

Adding the same number:

+4 +4 +4 +4 +4 +4 +4 +4
20, 24, 28, 32, 36, 40, 44, 48, 52

Doubling:

+5 +10 +20 +40 +80 +160 +320
5, 10, 20, 40, 80, 160, 320, 640

Patterns in the numbers being added:

+1 +2 +3 +4 +5 +6 +7 +8 +9
1, 2, 4, 7, 11, 16, 22, 29, 37, 46

PRACTICE | Fill the blanks to complete each pattern below.

52. 2, 4, 6, 8, 10, _____, _____, _____, _____, _____, _____

53. 21, 23, 25, 27, 29, _____, _____, _____, _____, _____, _____

54. 19, 29, 39, 49, _____, _____, _____, _____, _____, _____

55. 1, 2, 4, 8, 16, _____, _____, _____, _____, _____

PRACTICE | Fill the blanks to complete each pattern below.

56. 18, 21, 24, 27, 30, _____, _____, _____, _____, _____

57. 15, 16, 18, 21, 25, _____, _____, _____, _____, _____

58. 13, 24, 35, 46, 57, _____, _____, _____, _____, _____

59. 9, 11, 10, 12, 11, 13, 12, _____, _____, _____, _____

60. 25, 50, 100, 200, 400, _____, _____, _____, _____

61. 17, 18, 20, 21, 23, 24, 26, _____, _____, _____, _____

62. 1, 3, 7, 13, 21, 31, 43, _____, _____, _____, _____
★

63. 2, 2, 4, 6, 10, 16, 26, _____, _____, _____, _____
★
★

EXAMPLE | Posts are placed every mile along a road. The road has 87 posts, including one at the beginning and one at the end. How many miles long is the road?

You might think that with 87 posts, the road must be 87 miles long. Trying a simpler problem makes it clear that this isn't right.

What if there are just 2 posts? Then, the first post is just 1 mile from the second, and the road is 1 mile long.

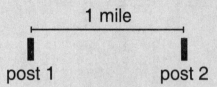

If there are 3 posts, the road is 2 miles long.

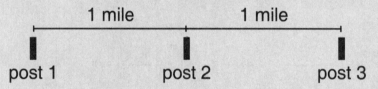

We can continue this pattern to see that each post after the first adds 1 mile to the length of the road.

Since there are 87 posts, the road is **86** miles long.

> Sometimes, the key to solving a hard problem is solving a similar but easier problem first.

PRACTICE | Solve each problem below.

64. Lizzie reads every page in a chapter that starts on page 6 and ends on page 8. How many pages does she read?

64. _____

65. Lizzie reads every page in a chapter that starts on page 25 and ends on page 30. How many pages does she read?

65. _____

66. ★ Lizzie reads every page in a chapter that starts on page 98 and ends on page 152. How many pages does she read?

66. _____

PRACTICE | Solve each problem below. The first has been solved for you. Look for a pattern that will help you solve the last problem.

Ex. There are 3 dots on the circle to the right. How many straight lines are needed to connect every dot to the other two dots?

Ex. __3__

67. There are 4 dots on the circle to the right. How many straight lines are needed to connect every dot to the other three dots?

67. _____

68. There are 5 dots on the circle to the right. How many straight lines are needed to connect every dot to the other four dots?

68. _____

69. There are 6 dots on the circle to the right. How many straight lines are needed to connect every dot to the other five dots?

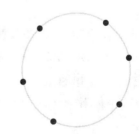

69. _____

70. ★ There are 7 dots on a circle. How many straight lines are needed to connect every dot to all of the other dots?

70. _____

PRACTICE | Solve each problem below. It may help to replace the underlined number with smaller numbers and look for a pattern.

71. Every tripticorn has 4 legs and 3 horns. How many *more* legs than horns are there in a barn that holds 26 tripticorns?

71. _____

72. How many cuts will it take Harry to split one carrot into 50 pieces if he cuts just one piece of carrot at a time?

72. _____

73. Minna folds a rectangular sheet of paper in half. Then, she folds the folded sheet in half again. She continues folding the sheet in half until she has folded it in half 7 times. When she unfolds the paper, the fold lines split the paper into small rectangles. How many small rectangles are there?

73. _____

PRACTICE | Solve each problem below. It may help to replace the underlined number with smaller numbers and look for a pattern.

74.   Find the sum of 10 copies of 74 .

74. _____

75.   When 4-inch cups are stacked in the Beast Academy cafeteria,
★   they overlap as shown. How many inches tall is a stack of 20
   cups stacked this way?

75. _____

1 inch

4 inches

3 inches

76.   Paulo arranges toothpicks to make a row of triangles like
★   the one below. How many toothpicks will he need to make
   a row of 25 triangles?

76. _____

EXAMPLE | Which of the following is the sum of 2,345 and 5,678?

6,789   7,925   8,023   9,123

We can answer this question without finding the actual sum by *eliminating* the answer choices that don't make sense.

We are adding a number between 2,000 and 3,000 to a number between 5,000 and 6,000. So, the sum is greater than 2,000+5,000 = 7,000, but less than 3,000+6,000 = 9,000.

So, the sum can't be 6,789 or 9,123.

~~6,789~~   7,925   8,023   ~~9,123~~

We can also find the ones digit of the sum without finding the whole sum. We just add the ones digits.

For many multiple choice questions, figuring out which choices are *wrong* can lead you to the answer that is right.

$$\begin{array}{r} 1 \\ 2,34\boxed{5} \\ +\ 5,67\boxed{8} \\ \hline \boxed{3} \end{array}$$

In the ones place, 5+8 = 13. So, the ones digit of the sum is 3. There is only one remaining answer choice with ones digit 3, so the sum must be **8,023**.

PRACTICE | Circle the correct answer for each problem below.

77. Find the sum of 45,561 and 19,871.

432   5,432   65,432   765,432   8,765,432

78. Find the sum of 148,327 and 98,537.

246,863   246,864   246,865   246,866   246,867

PRACTICE | Each problem below has one correct answer.
Circle the correct answer for each.

79. Which of the following is equal to 14,296+13,297+12,298?

    27,981   38,189   39,891   83,198   93,891

80. Find the sum of one hundred twenty-three 6's.

    731   733   735   737   738

81. Felix buys 10 boxes of crayons. Each box holds the same number of crayons. How many crayons could there be all together?

    317   318   319   320   321

82. Every stardvark has five snouts. How many total snouts could be on all of the stardvarks in the Beast Island Zoo?

    211   212   213   214   215

83. ★ What is 2+24+246+2,468+246,810+24,681,012?

    24,930,561   24,930,562   24,930,563   34,930,562   34,930,564

EXAMPLE

Three monsters weigh 10, 20, and 30 pounds. They are trying to cross a river using a boat that can carry up to 30 pounds. What is the smallest number of times the boat must cross the river to get all three monsters safely to the other side?

We can use labeled pieces of paper to stand for the monsters and the boat.

10 20 30

We can't start by sending one monster across alone, since that monster will just have to paddle back across to bring the boat back. So, we start by sending the 10- and 20-pound monsters across.

30   10 20 →

Then, one of these monsters must paddle the boat back.

30   ← 10   20

Sometimes, it helps to use objects to act out the problem.

The 30-pound monster can now cross alone.

10   30 →   20

The 20-pound monster can take the boat back.

10   ← 20   30

Finally, the 10- and 20-pound monsters can cross.

10 20 →   30

The boat crossed a total of 5 times. We could have switched the 10- and 20-pound monsters in the solution above, but **5** is the smallest number of trips possible.

PRACTICE | Find the smallest number of crossings needed to solve each river crossing problem below.

Find printable monsters to cut out at BeastAcademy.com.

84. One 30-pound monster and three 10-pound monsters are crossing a river using a boat that can carry up to 30 pounds. How many times will the boat need to cross the river?

84. _____

85. Four monsters weighing 10, 20, 30, and 40 pounds are crossing a river using a boat that can carry up to 50 pounds. The 10-pound monster is too young to paddle the boat or be left alone. How many times will the boat need to cross the river?

85. _____

86. ★ Two 10-pound monsters and two 20-pound monsters are crossing a river using a boat that can carry up to 20 pounds. How many times will the boat need to cross the river?

86. _____

87. ★ A farmer is taking a fox, a hen, and a bag of corn across a river using a boat. The farmer can only take one thing in the boat with her at a time. If she leaves the fox alone with the hen, the fox will eat the hen. If she leaves the hen alone with the corn, the hen will eat the corn. How many times will the boat need to cross the river?

87. _____

In a **Hopswitch** puzzle, the goal is to switch the positions of the black pawns and the white pawns on a grid of squares in as few moves as possible.

There are two ways that a pawn can move to an empty square:

By hopping over one pawn to the empty square on the other side.

— *or* —

By moving up, down, left, or right to an empty square.

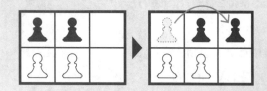

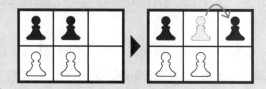

Pawns *cannot* move or hop diagonally, or hop over more than one pawn in a single hop. Each hop counts as one move.

EXAMPLE | What is the smallest number of moves needed to complete the Hopswitch puzzle below?

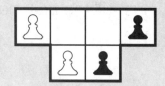

We can complete the puzzle with the moves shown below.

There are other ways to complete the puzzle in 8 moves, but **8** is the smallest number of moves possible.

Beast Academy Practice 2D

PRACTICE | Use pawns or other small items to help you find the smallest number of moves needed to complete each Hopswitch puzzle.

88. Best: _____ moves

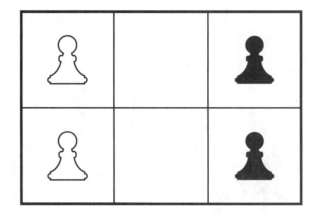

89. Best: _____ moves

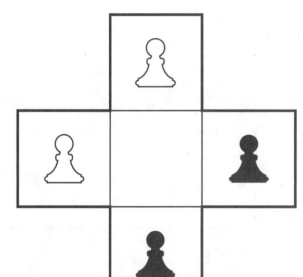

90. Best: _____ moves
★

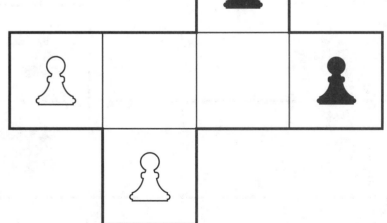

PRACTICE | Use pawns or other small items to help you find the smallest number of moves needed to complete each Hopswitch puzzle.

91. Good: 9 moves
★ Best: _____ moves

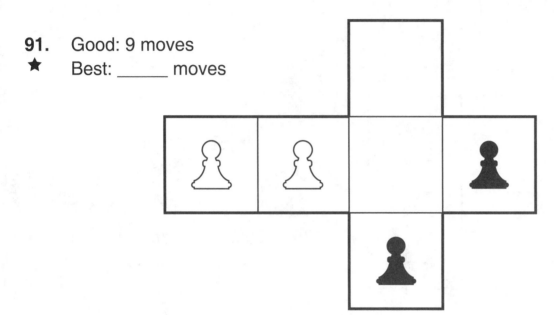

92. Good: 10 moves
★ Best: _____ moves

93. Good: 12 moves
★ Best: _____ moves

PRACTICE | Use pawns or other small items to help you find the smallest number of moves needed to complete each Hopswitch puzzle.

94. ★
Good: 12 moves
Better: 9 moves
Best: _____ moves

95. ★★
Good: 11 moves
Better: 9 moves
Best: _____ moves

96. ★★
Good: 12 moves
Better: 10 moves
Best: _____ moves

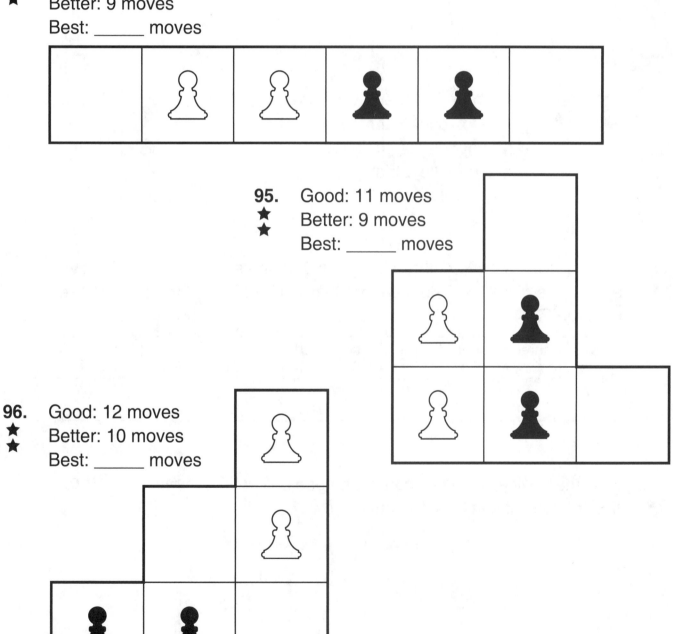

PRACTICE | Solve each problem below.

97. ★ How many different numbers between 700 and 1,000 have digits that add up to 10?

97. _____

98. ★ How many different ways are there to connect three of the four dots on the circle to the right to make a triangle?

98. _____

99. ★ Adding two 2's gives 4, three 3's gives 9, four 4's gives 16, five 5's gives 25, and six 6's gives 36. Find a pattern that will help you fill the empty boxes in the chart below.

two 2's	three 3's	four 4's	five 5's	six 6's	seven 7's	eight 8's	nine 9's	ten 10's	eleven 11's
4	9	16	25	36					

100. ★ How many different paths can the mouse below take to reach the cheese if it can only go right or down?

100. _____

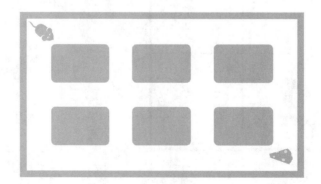

PRACTICE | Solve each problem below.

101. Show all ten ways to shade *three* of the triangles on the shape below.
★

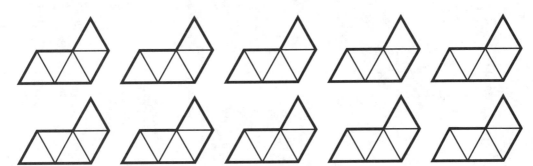

102. How many different ways can you use 1-cent pennies,
★
★ 5-cent nickels, and 10-cent dimes to make 25 cents?
For example, 10 pennies, 1 dime, and 1 nickel is one way
to make 25 cents.

102. _____

103. What is the smallest number of moves needed to
★
★ complete the Hopswitch puzzle below?

103. _____ moves

HINTS
For Selected Problems

Below are hints to every problem marked with a ★.
Work on the problems for a while before looking at the hints.
The hint numbers match the problem numbers.

CHAPTER 10
Big Numbers
6

30. After placing the 6-digit number, the 3-digit number, and the 4-digit number, where does fifty-five thousand fifty go?

31. The 5-digit number in the right column ends with the same digit as the 5-digit number in the bottom row. Which two 5-digit numbers end in the same digit?

38. Can you write the first three numbers using just 7 of the blanks?

39. Start by connecting the first number and the last number.

55. What number comes before 103,000? Before that?

57. What number comes before 1,011,000? How much more do you need to subtract?

67. How do you write 1 ten? 10 tens? 100 tens?

How do you write 1 hundred? 10 hundreds? 100 hundreds?

How do you write 1 thousand? 10 thousands? 100 thousands?

76. Add the thousands, then add the rest.

77. Add the thousands, then add the rest.

86. Take away the thousands, then take away the rest.

87. Take away the thousands, then take away the rest.

88. Is there a better way to write "20 hundreds"? What about "20 tens"?

89. Take away the millions, the thousands, and then the rest.

93. Add the millions, the thousands, and then the rest.

97. What is 500+800? What is 500 thousands plus 800 thousands?

98. What is 750+740? What is 750 thousands plus 740 thousands?

105. What is the largest number you can make? The smallest?

106. How much greater than 100,000 is 99,000,000?

118. Write each number using digits.

122. What is the ones digit of the number?

123. How many of the numbers are in the 1,200's? The 1,300's?

125. How many possibilities are there? Try them all.

126. Which digits are more important to the size of a number, the digits on the left, or on the right?

127. Which digits are more important to the size of a number, the digits on the left, or on the right?

135. The rightmost number has just three digits. What does this tell you about the digits you must cross out?

143. What two numbers that end in 0,000 is 764,971 between? What two numbers that end in 000 is 764,971 between? What two numbers that end in 00 is 764,971 between?

153. Compare both sums to 1,200.

154. Compare both sums to 15,000.

157. What is the smallest number? What is a good way to subtract it?

158. What are the three largest 5-digit numbers? What is a good way to add them?

159. How much does switching the digits increase the thousands? How much does it decrease the ones?

161. What is the closest number you can write that is in the 4,000's? What is the closest number you can write that is in the 5,000's? Which of these is closer?

162. What is a better way to write "10 hundreds"? What about "10 tens"?

163. A 7-digit number is in the millions. What digit should be in the millions place? The ones place?

CHAPTER 11
Algorithms
38

45. What digit must be in the thousands place of the top number to create a 5-digit sum?

46. What digit must be in the ones place of the top number to make the tens digit of the sum 1?

62. In the tens column, either $2+\boxed{0}=\boxed{2}$ or $2+\boxed{8}=1\boxed{0}$. Which of these works?

63. You don't always have to work from right to left. Which of the given digits could fill the blank in the thousands column? The hundreds column?

64. Look at the sum in the hundreds column. What does this tell us about the sum in the tens column?

75. After you find the missing ones digits of the left and right sums, which of the remaining digits can be the missing digit in the hundreds column of the right sum (between the 2 and the 9)?

76. After finding the missing ones digit of the right sum, which of the remaining digits can be the missing hundreds digits?

77. Which of the remaining digits can be the missing digit in the hundreds column of the left sum?

78. Which two of the remaining digits can fill the blanks in the ones column of the left sum?

Which two of the remaining digits can fill the blanks in the ones column of the right sum?

130. How many cards are in Borg's carrying case?

137. In the ones column, A+B ends in A. What digit is B?

138. What digit is D? Then, what are the B's in the hundreds column?

140. What numbers could we subtract to find the missing number in an easier problem like 56 − ___ = 11?

142. How much greater is the second sum than the first sum?

143. Which digits can be used to fill both blanks in the ones column?

144. Try adding some numbers that only use 5's as digits. What do you notice?

145. To make the difference as small as possible, the thousands digits should be as close as possible. What else could you do to make the numbers as close together as possible?

146. Stack the addition. The ten-thousands digit of the sum is A. What does this tell you about A?

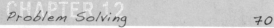

CHAPTER 12
Problem Solving   70

9. There are small triangles and large triangles. How many of the large triangles point to the left? How many point right?

16. There are 4 paths that start by going exactly one block down.

23. There are 4 paths that start by going down.

24. There are 4 paths that start by going left.

37. How many ways can you shade the top square plus one square below it? How many ways can you shade the second square from the top plus one square below it? Keep working your way down.

38. How many of the shadings you made have an upside-down "twin"?

39. Instead of thinking about which squares *are* shaded, it may help to think about which squares are *not* shaded.

45. Which digits can be the thousands digit?

46. How many have a 1 in the thousands place? List them from least to greatest.

51. How many sums can you make that use two 3's? One 3? No 3's?

62. What is the pattern in the numbers you are adding?

63. Take a look at the numbers you are adding. Where have you seen those numbers before?

66. In the two previous problems, does subtracting the page numbers give you the right answer?

70. Find a pattern in the previous problems that will help you figure out how many lines are needed.

75. How tall is a stack of 2 cups? 3? 4? 20? 200?

76. How many toothpicks does it take to make 1 triangle? A row of two triangles? 3? 4? 10? 25?

83. Is the sum odd or even? About how big is the sum?

86. Who should cross the river together first?

87. Which item can the farmer take across first? (Who can she leave alone on the shore together?)

90. How many moves does it take to solve this Hopswitch puzzle?

Can you solve it twice?

91. Start with this move:

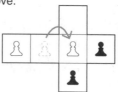

Finish by jumping one pawn over another.

92. Try to avoid placing two pawns of the same color next to each other, since this blocks pawns of the other color from jumping over them.

93. Make your first two moves with the leftmost pawn.

94. Can you get to either of the positions below in four moves?

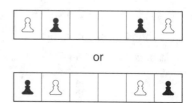

or

95. You don't even need to use one of the squares in this puzzle!

96. Start with this move:

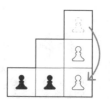

Look for ways to save moves by jumping pawns over each other instead of moving one square at a time.

97. How many of these numbers are in the 700's? The 800's? The 900's?

98. When you connect three dots to make a triangle, one dot is left out.

99. How much do you add to get from 4 to 9? From 9 to 16? From 16 to 25?

100. If the mouse starts by going all the way to the right, how many ways are there to reach the cheese? What if it starts by going two "blocks" to the right? One block to the right? No blocks to the right?

101. It may be easier to organize your work if you think about which two triangles are left unshaded.

102. Organize your work by the number of dimes, then by the number of nickels.

103. Try to avoid placing two pawns of the same color next to each other, since this blocks pawns of the other color from jumping over them.

SOLUTIONS
Chapters 10-12

BIG NUMBERS

Review 7

1. We circle each number with tens digit 8 as shown.

18   (8̲1)   118   (18̲8)   818   (88̲1)

2. We circle each number whose ones digit is the same as its hundreds digit as shown.

(7̲37)   337   733   (3̲73)   733   377

3. We circle each number whose ones digit is greater than its tens digit as shown.

92   (229)   292   922   992   (29)

4. We begin by writing 4 in the tens place.

_ 4 _

The remaining digits are 0 and 2. We cannot write 0 in the hundreds place, so we write 2 in the hundreds place and 0 in the ones place.

2 4 0

5. The hundreds digit is greater than the ones digit, and the ones digit is greater than the tens digit. So, the hundreds digit is the largest digit, 7. The ones digit is the next-largest digit, 6. Finally, the tens digit is the smallest digit, 5.

So, the three-digit number is **756**.

BIG NUMBERS

Larger Place Values 8-9

6. We circle each ten-thousands digit as shown.

8̲7,209   1̲5,426   37̲8,921   35̲6,278

7. We circle each thousands digit as shown.

42̲,809   75̲,021   529̲,481   123̲,456

8. We circle each number with hundred-thousands digit 7 as shown.

78,786   (725,325)   177,771   (707,070)

9. We circle each ten-millions digit as shown.

17̲4,870,932   8̲9,647,150   70̲6,813,924   3̲0,458,972

10. We circle each number whose millions digit is the same as its ten-thousands digit as shown.

(808,080,808)   123,123,123   (234,543,210)   567,765,567

11. We circle each number whose tens digit is greater than its ten-millions digit as shown.

(123,456,789)   978,675,645   232,121,434   (624,910,538)

BIG NUMBERS

Reading Numbers 10-11

We connect each number on the left with its matching number on the right as shown.

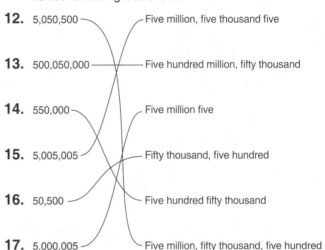

12. 5,050,500 — Five million, five thousand five

13. 500,050,000 — Five hundred million, fifty thousand

14. 550,000 — Five million five

15. 5,005,005 — Fifty thousand, five hundred

16. 50,500 — Five hundred fifty thousand

17. 5,000,005 — Five million, fifty thousand, five hundred

18. We write the number of thousands, then the rest. There are 345 thousands and 986.

Three hundred forty-five thousand,   nine hundred eighty-six.

(345),(986)

So, we have **345,986**.

19. We write the number of millions, then thousands, then the rest. There are 78 millions, 340 thousands, and nothing else.

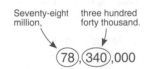

Seventy-eight million,   three hundred forty thousand.

(78),(340),000

So, we have **78,340,000**.

20. We write the number of millions, then thousands, then the rest. There are 406 millions, no thousands, and 972.

Four hundred six million,   nine hundred seventy-two.

(406),000,(972)

So, we have **406,000,972**.

21. We write the number of millions, then thousands, then the rest. There are 30 millions, 7 thousands, and nothing else.

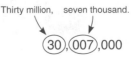

Thirty million,   seven thousand.

(30),(007),000

So, we have **30,007,000**.

22. We write the number of millions, then thousands, then the rest. There are 700 millions, 24 thousands, and 638.

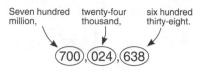

So, we have **700,024,638**.

23. We write the number of millions, then thousands, then the rest. There are 59 millions, no thousands, and 59.

Fifty-nine million   fifty-nine.

(59),000,(059)

So, we have **59,000,059**.

24. We write the number of thousands, then the rest. There are 212 thousands and 212.

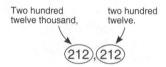

So, we have **212,212**.

25. We write the number of millions, then thousands, then the rest. There are 802 millions, 2 thousands, and 2.

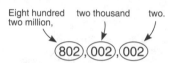

So, we have **802,002,002**.

BIG NUMBERS
Numbercross   12–13

26. We write each number using digits.

Nine hundred nine = 909.
Nine hundred ninety = 990.
Nine thousand ninety-nine = 9,099.
Ninety-nine thousand ninety = 99,090.

9,099 is the only four-digit number and 99,090 is the only five-digit number. So, we place these numbers as shown.

9				
0				
9	9	0	9	0
9				

The remaining numbers are 909 and 990. There is only one way to place these two numbers, as shown.

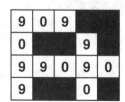

27. We write each number using digits.

Seven hundred seventy = 770.
Seven thousand seventy = 7,070.
Seven thousand seven hundred = 7,700.
Seventy-seven thousand seven = 77,007.

770 is the only three-digit number and 77,007 is the only five-digit number. So, we place these numbers as shown.

				7
7	7	0	0	7
				0

The remaining numbers are 7,070 and 7,700. There is only one way to place these two numbers, as shown.

7	7	0	0	
	0			7
7	7	0	0	7
	0			0

28. We write each number using digits.

One thousand two = 1,002.
Nine thousand seventy = 9,070.
Ten thousand, six hundred four = 10,604.
Twenty thousand, five hundred three = 20,503.
Eighty-four thousand three = 84,003.
Seven hundred six thousand five = 706,005.

We use the strategies discussed in previous problems to place these numbers as shown.

9		1	0	0	2
0		0			0
7	0	6	0	0	5
0		0			0
	8	4	0	0	3

29. We write each number using digits.

Three hundred seventy-five = 375.
Three thousand seventy-five = 3,075.
Fifty thousand, three hundred seven = 50,307.
Seventy-five thousand three = 75,003.
Three hundred seventy thousand, five hundred five = 370,505.
Five hundred thirty thousand, three hundred = 530,300.

We use the strategies discussed in previous problems to place these numbers as shown.

7	5	0	0	3	
	0				3
5	3	0	3	0	0
	0		7		7
3	7	0	5	0	5

30. We write each number using digits.

Three hundred three = 303.
Four thousand four = 4,004.
Thirty thousand three hundred = 30,300.
Fifty thousand, three hundred thirty = 50,330.
Fifty-five thousand fifty = 55,050.
Six hundred thousand, three hundred four = 600,304.

We begin by placing 303, 4,004, and 600,304 as shown.

					■
■					**4**
■			**3**		**0**
		■	**0**	■	**0**
6	**0**	**0**	**3**	**0**	**4**

We cannot place 55,050 in the second column, since the number in the third row cannot start with 0.

	5				■
■	**5**	■			**4**
■	**0**		**3**		**0**
	5	■	**0**	■	**0**
6	**0**	**0**	**3**	**0**	**4**

✗

So, we can only place 55,050 in the top row.

5	**5**	**0**	**5**	**0**	■
■		■			**4**
■			**3**		**0**
		■	**0**	■	**0**
6	**0**	**0**	**3**	**0**	**4**

Then, we complete the puzzle as shown.

5	**5**	**0**	**5**	**0**	■
■	**0**	■			**4**
■	**3**	**0**	**3**	**0**	**0**
	3	■	**0**	■	**0**
6	**0**	**0**	**3**	**0**	**4**

31. We write each number using digits.

Four hundred forty = 440.
Four thousand four = 4,004.
Four thousand four hundred = 4,400.
Forty thousand forty = 40,040.
Forty thousand, four hundred four = 40,404.
Forty-four thousand, four hundred = 44,400.

We begin by placing 440 as shown.

	■			■	
4	**4**	**0**	■		
	■		■		
	■		■		
■			■		

Then, there is only one way to place 4,004 and 4,400.

4	■	**4**	**0**	**0**	**4**
4	**4**	**0**	■		
0	■		■		
0	■		■		
■			■		

The numbers in the right column and the bottom row have the same last digit. Among the remaining numbers, only 40,04<u>0</u> and 44,40<u>0</u> have the same last digit, so these numbers go in the right column and bottom row. The only other remaining number is 40,404. So, we place 40,404 in the third column.

4	■	**4**	**0**	**0**	**4**
4	**4**	**0**	■		
0	■	**4**	■		
0	■	**0**	■		
■		**4**			

Finally, we complete the puzzle as shown.

4	■	**4**	**0**	**0**	**4**
4	**4**	**0**	■		**0**
0	■	**4**	■		**0**
0	■	**0**	■		**4**
■	**4**	**4**	**4**	**0**	**0**

32. First, we write each number using digits.

One hundred two = 102.
Two hundred three = 203.
Three hundred four = 304.

Then, we fill the blanks so that each number appears in the row.

<u>**1 0 2 0 3 0 4**</u>

One hundred two	**1 0 2** 0 3 0 4
Two hundred three	1 0 **2 0 3** 0 4
Three hundred four	1 0 2 0 **3 0 4**

33. We write each number using digits.

Five thousand seventy = 5,070.
Five thousand, five hundred seven = 5,507.
Six thousand fifty-five = 6,055.

We can include 5,507 and 6,055 by writing 605507.
Then, we write 0 at the end of 605507 to include 5,070.

6 0 5 5 0 7 0

Five thousand seventy	6 0 5 **5 0 7 0**
Five thousand five hundred seven	6 0 **5 5 0 7** 0
Six thousand fifty-five	**6 0 5 5** 0 7 0

34. We write each number using digits.

One thousand twenty-three = 1,023.
Three thousand twenty-one = 3,021.
Two hundred ten = 210.

We can include 3,021 and 210 by writing 30210. Then, we write 23 at the end of 30210 to include 1,023.

3 0 2 1 0 2 3

One thousand twenty-three	3 0 2 1 **0 2 3**
Three thousand twenty-one	**3 0 2 1** 0 2 3
Two hundred ten	3 0 **2 1 0** 2 3

35. We write each number using digits.

One hundred thirty-one = 131.
One thousand thirty-one = 1,031.
Three thousand, one hundred thirteen = 3,113.

We can include 1,031 and 3,113 by writing 103113.
Then, we write 1 at the end of 103113 to include 131.

1 0 3 1 1 3 1

One hundred thirty-one	1 0 3 1 **1 3 1**
One thousand thirty-one	**1 0 3 1** 1 3 1
Three thousand, one hundred thirteen	1 0 **3 1 1 3** 1

36. We write each number using digits.

Twenty thousand twelve = 20,012.
One thousand two hundred = 1,200.
Two hundred two = 202.

There are two ways to connect 20,012 and 1,200 so that some of their digits overlap:

2001200 **or** 120012.

There are only eight blanks given. We cannot include 202 in 2001200 using eight digits. So, we write 02 at the end of 120012 to include 202.

1 2 0 0 1 2 0 2

Twenty thousand twelve	**1 2 0 0 1 2** 0 2
One thousand two hundred	**1 2 0 0** 1 2 0 2
Two hundred two	1 2 0 0 1 **2 0 2**

37. We write each number using digits.

Five hundred six = 506.
Four thousand fifty = 4,050.
Five thousand forty = 5,040.
Six thousand fifty = 6,050.

To connect 4,050, 5,040, and 6,050 using the fewest number of digits, we write 60504050. Then, we write 6 at the end of 60504050 to include 506.

6 0 5 0 4 0 5 0 6

Five hundred six	6 0 5 0 4 0 **5 0 6**
Four thousand fifty	6 0 5 0 **4 0 5 0** 6
Five thousand forty	6 0 **5 0 4 0** 5 0 6
Six thousand fifty	**6 0 5 0** 4 0 5 0 6

38. We write each number using digits.

Seven thousand, five hundred seven = 7,507.
Five thousand seventy-five = 5,075.
Five thousand seventy-seven = 5,077.
Seven thousand seventy-seven = 7,077.

7,507 starts with 75 and 5,075 ends with 75. So, we can connect these two numbers by writing 507507.

Then, we write 7 at the end of 507507 to include 5,077.

5075077 is seven digits long, and there are ten blanks given. So, we can only add three digits to include 7,077. Only by writing 077 at the end of 5075077 can we include 7,077.

5 0 7 5 0 7 7 0 7 7

Seven thousand, five hundred seven	5 0 **7 5 0 7** 7 0 7 7
Five thousand seventy-five	**5 0 7 5** 0 7 7 0 7 7
Five thousand seventy-seven	5 0 7 **5 0 7 7** 0 7 7
Seven thousand seventy-seven	5 0 7 5 0 7 **7 0 7 7**

39. We write each number using digits.

Nine thousand, four hundred forty-nine = 9,449.
Four thousand forty-four = 4,044.
Four thousand, four hundred ninety-four = 4,494.
Forty-nine thousand forty-four = 49,044.

4,494 starts with 44. Both 4,044 and 49,044 end with 44.

We try connecting 4,494 with 49,044 by writing 4904494. 4904494 is seven digits long, and there are eleven blanks given. So, we can only add four digits to 4904494. But, it is impossible to include both 4,044 and 9,449 by adding four digits to 4904494. ✘

So, we try connecting 4,494 with 4,044 by writing 404494. Then, we write 49 at the end of 404494 to include 9,449.

40449449 is eight digits long, so we can only add three digits to 40449449 to include 49,044. This is only possible if we write 044 at the end of 40449449.

4 0 4 4 9 4 4 9 0 4 4

Nine thousand, four hundred forty-nine	4 0 4 4 **9 4 4 9** 0 4 4
Four thousand forty-four	**4 0 4 4** 9 4 4 9 0 4 4
Four thousand, four hundred ninety-four	4 0 **4 4 9 4** 4 9 0 4 4
Forty-nine thousand forty-four	4 0 4 4 9 **4 4 9 0 4 4**

40. Counting up from 996 we have 997, 998, then 999. Then, one more than 999 is 1,000.

<u>995</u> <u>996</u> **997** **998** **999** **1,000** <u>1,001</u>

41. Counting up from 179,997 we have 179,998, then 179,999. Then, one more than 179,999 is 180,000.

<u>179,997</u> **179,998** **179,999** **180,000** <u>180,001</u>

42. Counting down from 84,997 we have 84,996, then 84,995. Counting up from 84,998 we have 84,999, then 85,000.

84,995 **84,996** <u>84,997</u> <u>84,998</u> **84,999** **85,000**

43. One more than 1,099,999 is 1,100,000. Then, one more than 1,100,000 is 1,100,001.

<u>1,099,998</u> <u>1,099,999</u> **1,100,000** **1,100,001**

44. Counting up from 3,899,998 we have 3,899,999, then 3,900,000.

<u>3,899,997</u> <u>3,899,998</u> **3,899,999** **3,900,000**

45. We count down from 100,002 until we get to 100,000. Then, one less than 100,000 is 99,999, and one less than 99,999 is 99,998.

99,998 **99,999** **100,000** **100,001** <u>100,002</u> <u>100,003</u>

46. We count up by 3 from 99,999.

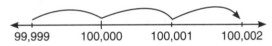

So, 3 more than 99,999 is **100,002**.

— *or* —

To find 3 more than 99,999, we add 99,999+3.

To add 3 to 99,999, we add 1 then add 2. 99,999+1 = 100,000. Then, 100,000+2 = **100,002**.

47. We count down by 4 from 10,001.

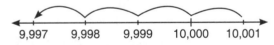

So, 4 less than 10,001 is **9,997**.

— *or* —

To find 4 less than 10,001, we subtract 10,001−4.

To subtract 4 from 10,001, we subtract 1 then subtract 3. 10,001−1 = 10,000. Then, 10,000−3 = **9,997.**

48. We count up by 3 from 1,998.

So, 1,998+3 = **2,001.**

— *or* —

To add 3 to 1,998, we add 2 then add 1. 1,998+2 = 2,000. Then, 2,000+1 = **2,001.**

49. We count down by 5 from 140,002.

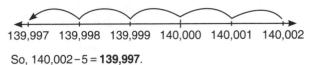

So, 140,002−5 = **139,997**.

— *or* —

To subtract 5 from 140,002, we subtract 2 then subtract 3. 140,002−2 = 140,000. Then, 140,000−3 = **139,997**.

50. To add 7 to 99,996, we add 4 then add 3. 99,996+4 = 100,000. Then, 100,000+3 = **100,003**.

51. To subtract 6 from 1,000,004, we subtract 4 then subtract 2. 1,000,004−4 = 1,000,000. Then, 1,000,000−2 = **999,998**.

52. To add 11 to 31,992, we add 8 then add 3. 31,992+8 = 32,000. Then, 32,000+3 = **32,003**.

53. To subtract 9 from 45,002, we subtract 2 then subtract 7. 45,002−2 = 45,000. Then, 45,000−7 = **44,993**.

54. To add 12 to 49,996, we add 4 then add 8. 49,996+4 = 50,000. Then, 50,000+8 = **50,008**.

55. To subtract 4 from 103,002, we subtract 2 then subtract 2. 103,002−2 = 103,000. Then, 103,000−2 = **102,998**.

56. To add 6 to 9,099,099, we add 1 then add 5. 9,099,099+1 = 9,099,100. Then, 9,099,100+5 = **9,099,105**.

57. To subtract 7 from 1,011,001, we subtract 1 then subtract 6. 1,011,001−1 = 1,011,000. Then, 1,011,000−6 = **1,010,994**.

58. We count by 10's to complete the list as shown.

<u>950</u> <u>960</u> **970** **980** **990** **1,000** <u>1,010</u>

Notice that counting by 10's is like counting by 1's but with a 0 at the end.

59. We count by 1,000's to complete the list as shown.

<u>196,000</u> <u>197,000</u> **198,000** **199,000** **200,000**

Notice that counting by 1,000's is like counting by 1's but with three 0's at the end.

60. We count by 100's to complete the list as shown.

<u>39,600</u> <u>39,700</u> **39,800** **39,900** **40,000**

Notice that counting by 100's is like counting by 1's but with two 0's at the end.

61. We count by 10's to complete the list as shown.

<u>109,970</u> <u>109,980</u> **109,990** **110,000** **110,010**

Notice that counting by 10's is like counting by 1's but with a 0 at the end.

62. We count by 100's to complete the list as shown.

<u>89,800</u> <u>89,900</u> **90,000** **90,100** **90,200**

Notice that counting by 100's is like counting by 1's but with two 0's at the end.

63. We count by 1,000's to complete the list as shown.

<u>1,098,000</u> <u>1,099,000</u> **1,100,000** **1,101,000** **1,102,000**

Notice that counting by 1,000's is like counting by 1's but with three 0's at the end.

64. 1 hundred is 100, 2 hundreds is 200, 3 hundreds is 300, and so on. Any number of hundreds is just that number followed by two 0's. So, we fill the blanks as shown.

8 hundreds = <u>800</u> 10 hundreds = **<u>1,000</u>**

13 hundreds = **<u>1,300</u>** 17 hundreds = **<u>1,700</u>**

20 hundreds = **<u>2,000</u>** 100 hundreds = **<u>10,000</u>**

65. 1 ten is 10, 2 tens is 20, 3 tens is 30, and so on. Any number of tens is just that number followed by one 0. So, we fill the blanks as shown.

5 tens = **<u>50</u>** 9 tens = **<u>90</u>**

10 tens = **<u>100</u>** 13 tens = **<u>130</u>**

40 tens = **<u>400</u>** 555 tens = **<u>5,550</u>**

66. 1 thousand is 1,000, 2 thousands is 2,000, 3 thousands is 3,000, and so on. Any number of thousands is just that number followed by three 0's. So, we fill the blanks as shown.

7 thousands = **<u>7,000</u>** 70 thousands = **<u>70,000</u>**

700 thousands = **<u>700,000</u>** 4,000 thousands = **<u>4,000,000</u>**

67. Any number of tens is just that number followed by one 0.

Any number of hundreds is just that number followed by two 0's.

Any number of thousands is just that number followed by three 0's.

So, we fill the blanks as shown.

100,000 tens = **<u>1,000,000</u>**

100,000 hundreds = **<u>10,000,000</u>**

100,000 thousands = **<u>100,000,000</u>**

BIG NUMBERS
Addition & Subtraction 20-21

68. 12 thousands plus 1 thousand is
$12+1 = 13$ thousands, or **13,000**.

— *or* —

Adding 1,000 to 12,000 increases the thousands digit by 1. So, $12,000+1,000 =$ **13,000**.

69. 41 thousands plus 6 thousands is
$41+6 = 47$ thousands, or **47,000**.

— *or* —

Adding 6,000 to 41,000 increases the thousands digit by 6. So, $41,000+6,000 =$ **47,000**.

70. 333 thousands plus 10 thousands is
$333+10 = 343$ thousands, or **343,000**.

— *or* —

Adding 10,000 to 333,000 increases the ten-thousands digit by 1. So, $333,000+10,000 =$ **343,000**.

71. 35 thousands plus 40 thousands is
$35+40 = 75$ thousands, or **75,000**.

— *or* —

Adding 40,000 to 35,000 increases the ten-thousands digit by 4. So, $35,000+40,000 =$ **75,000**.

72. 650 thousands plus 300 thousands is
$650+300 = 950$ thousands, or **950,000**.

— *or* —

Adding 300,000 to 650,000 increases the hundred-thousands digit by 3. So, $650,000+300,000 =$ **950,000**.

73. Adding 20,000 to 4,567,000 increases the ten-thousands digit by 2. So, $4,567,000+20,000 =$ **4,587,000**.

74. Adding 80,000 to 713,975 increases the ten-thousands digit by 8. So, $713,975+80,000 =$ **793,975**.

75. Adding 6,000 to 92,756 increases the thousands digit by 6. So, $92,756+6,000 =$ **98,756**.

76. Adding 30,070 to 154,103 increases the ten-thousands digit by 3 and the tens digit by 7.

So, $154,103+30,070 =$ **184,173**.

77. Adding 500,600 to 375,232 increases the hundred-thousands digit by 5 and the hundreds digit by 6.

So, $375,232+500,600 =$ **875,832**.

78. 35 thousands minus 1 thousand is
$35-1 = 34$ thousands, or **34,000**.

— *or* —

Subtracting 1,000 from 35,000 decreases the thousands digit by 1. So, $35,000-1,000 =$ **34,000**.

79. 247 thousands minus 5 thousands is
$247-5 = 242$ thousands, or **242,000**.

— *or* —

Subtracting 5,000 from 247,000 decreases the thousands digit by 5. So, $247,000-5,000 =$ **242,000**.

80. 290 thousands minus 10 thousands is
$290-10 = 280$ thousands, or **280,000**.

— *or* —

Subtracting 10,000 from 290,000 decreases the ten-thousands digit by 1. So, $290,000-10,000 =$ **280,000**.

81. 450 thousands minus 30 thousands is
$450-30 = 420$ thousands, or **420,000**.

— *or* —

Subtracting 30,000 from 450,000 decreases the ten-thousands digit by 3. So, $450,000-30,000 =$ **420,000**.

82. 980 thousands minus 600 thousands is
980−600 = 380 thousands, or **380,000**.

— *or* —

Subtracting 600,000 from 980,000 decreases the
hundred-thousands digit by 6. So,
980,000−600,000 = **380,000**.

83. Subtracting 70,000 from 1,290,000 decreases the
ten-thousands digit by 7. So,
1,290,000−70,000 = **1,220,000**.

84. Subtracting 5,000 from 239,654 decreases the
thousands digit by 5. So, 239,654−5,000 = **234,654**.

85. Subtracting 60,000 from 78,987 decreases the
ten-thousands digit by 6. So, 78,987−60,000 = **18,987**.

86. Subtracting 200,300 from 756,492 decreases the
hundred-thousands digit by 2 and the hundreds digit by 3.

So, 756,492−200,300 = **556,192**.

87. Subtracting 80,020 from 796,562 decreases the
ten-thousands digit by 8 and the tens digit by 2.

So, 796,562−80,020 = **716,542**.

88. 20 thousands is 20,000.
20 hundreds is 2,000.
20 tens is 200.

So, we are looking for 20,000−2,000−200.

20,000−2,000 = 18,000, and 18,000−200 = **17,800**.

89. We subtract the millions, then the thousands, then the rest.

9,090,900 has 9 millions. Subtracting 5 millions gives
9−5 = 4 millions.

9,090,900 has 90 thousands. Subtracting 5 thousands
gives 90−5 = 85 thousands.

Finally, we subtract 5 tens from 900. This gives
900−50 = 850.

4 millions, 85 thousands, and 850 is **4,085,850**.

BIG NUMBERS
Adding in Groups   22-23

90. Adding the thousands gives 11+55 = 66 thousands, or
66,000. Adding the rest gives 250+75 = 325.

66,000+325 = 66,325. So, we fill the blanks as shown.

11,250+55,075 = $\boxed{6\,6},\boxed{0\,0\,0}=\boxed{6\,6},\boxed{3\,2\,5}$.

91. Adding the thousands gives 80+45 = 125 thousands, or
125,000. Adding the rest gives 90+70 = 160.

125,000+160 = 125,160. So, we fill the blanks as shown.

80,090+45,070 = $\boxed{1\,2\,5},\boxed{0\,0\,0}+\boxed{1\,6\,0}=\boxed{1\,2\,5},\boxed{1\,6\,0}$.

92. Adding the thousands gives 246+98 = 344 thousands, or
344,000. Adding the rest gives 392+208 = 600.

344,000+600 = 344,600. So, we fill the blanks as shown.

246,392+98,208 = $\boxed{3\,4\,4},\boxed{0\,0\,0}+\boxed{6\,0\,0}=\boxed{3\,4\,4},\boxed{6\,0\,0}$.

93. Adding the millions gives 43+35 = 78 millions, or
78,000,000.

Adding the thousands gives 480+70 = 550 thousands, or
550,000.

Adding the rest gives 278+300 = 578.

78,000,000+550,000+578 = 78,550,578. So, we fill the
blanks as shown.

43,480,278+35,070,300 = $\boxed{7\,8},\boxed{5\,5\,0},\boxed{5\,7\,8}$.

94. Adding the thousands gives 3+4 = 7 thousands, or 7,000.
Adding the rest gives 900+700 = 1,600.

7,000+1,600 = 8,600. So, we fill the blanks as shown.

3,900+4,700 = $\boxed{7},\boxed{0\,0\,0}+\boxed{1},\boxed{6\,0\,0}=\boxed{8},\boxed{6\,0\,0}$.

95. Adding the thousands gives 800+300 = 1,100 thousands,
or 1,100,000. So, we fill the blanks as shown.

800,000+300,000 = $\boxed{1},\boxed{1\,0\,0},\boxed{0\,0\,0}$.

96. Adding the thousands gives 30+50 = 80 thousands, or
80,000. Adding the rest gives 400+700 = 1,100.

80,000+1,100 = 81,100. So, we fill the blanks as shown.

30,400+50,700 = $\boxed{8\,0},\boxed{0\,0\,0}+\boxed{1},\boxed{1\,0\,0}=\boxed{8\,1},\boxed{1\,0\,0}$.

97. Adding the thousands gives 500+800 = 1,300 thousands,
or 1,300,000. Adding the rest gives 500+800 = 1,300.

1,300,000+1,300 = 1,301,300. So, we fill the blanks as
shown.

500,500+800,800 = $\boxed{1},\boxed{3\,0\,1},\boxed{3\,0\,0}$.

98. Adding the thousands gives 750+740 = 1,490 thousands,
or 1,490,000. Adding the rest gives 600+900 = 1,500.

1,490,000+1,500 = 1,491,500. So, we fill the blanks as
shown.

750,600+740,900 = $\boxed{1},\boxed{4\,9\,1},\boxed{5\,0\,0}$.

BIG NUMBERS
Word Problems   24-25

99. One bridge uses 80,000 miles of cable. So, two bridges
use 80,000+80,000 = **160,000** miles of cable.

100. A snow yeti has about 1,900,000+500,000 hairs.

To add 500,000 to 1,900,000, we add 100,000 then add
400,000.

1,900,000+100,000 = 2,000,000. Then,
2,000,000+400,000 = **2,400,000**.

101. Using the given estimates, there are 1,580,000−1,250,000
more tons of sand at Coral Cove than at Blowfish Beach.

To subtract 1,250,000 from 1,580,000, we subtract
1,000,000 then subtract 250,000.

1,580,000−1,000,000 = 580,000. Then,
580,000−250,000 = **330,000**.

102. The three smallest 5-digit numbers are 10,000, 10,001,
and 10,002. So, the sum of the three smallest 5-digit
numbers is 10,000+10,001+10,002 = 30,000+3 = **30,003**.

103. There are 5,280+5,280 feet in two miles.

Adding the thousands gives 5+5 = 10 thousands, or
10,000. Adding the rest gives 280+280 = 560.

So, 5,280+5,280 = 10,000+560 = **10,560**.

104. A week is 7 days. So, 8 days is 1 more day than a week. So, there are 604,800+86,400 seconds in 8 days.

Adding the thousands gives 604+86 = 690 thousands, or 690,000. Adding the rest gives 800+400 = 1,200.

So, 604,800+86,400 = 690,000+1,200 = **691,200**.

105. The largest number we can make with three 3's and three 0's is 333,000.

The smallest number we can make with three 3's and three 0's is 300,033.

Adding these numbers, we have
333,000+300,033 = **633,033**.

106. To find how much greater 99,000,000 is than 99,000, we subtract. So, we are looking for 99,000,000−99,000.

To subtract 99,000, we can subtract 100,000 then add 1,000 back. So, we have

$$99,000,000-99,000 = 99,000,000-100,000+1,000$$
$$= 98,900,000+1,000$$
$$= \mathbf{98,901,000.}$$

BIG NUMBERS
Comparing & Ordering
26-29

107. Any five-digit number with ten-thousands digit 2 is less than a five-digit number with ten-thousands digit 3. So, 23,232 is less than 32,323.

23,232 $<$ 32,323

108. 98,765 has fewer digits than 456,789. So, 98,765 is less than 456,789.

98,765 $<$ 456,789

109. 101,010 has more digits than 98,989. So, 101,010 is greater than 98,989.

101,010 $>$ 98,989

110. Both numbers have the same number of digits and the same first two digits. Since 81,200 is greater than 79,900, we know 2,181,200 is greater than 2,179,900.

2,181,200 $>$ 2,179,900

111. Both numbers have the same number of digits and the same first two digits. Since 678 is less than 876, we know 45,678 is less than 45,876.

45,678 $<$ 45,876

112. 9,009,009 has fewer digits than 10,001,001. So, 9,009,009 is less than 10,001,001.

9,009,009 $<$ 10,001,001

113. 777,777 has more digits than 88,888. So, 777,777 is greater than 88,888.

777,777 $>$ 88,888

114. Any seven-digit number with millions digit 2 is greater than a seven-digit number with millions digit 1. So, 2,000,001 is greater than 1,999,998.

2,000,001 $>$ 1,999,998

115. There is only one 5-digit number: 95,000.

There are two 6-digit numbers: 905,000 and 950,000. 905,000 is less than 950,000.

There is only one 7-digit number: 9,500,000.

There is only one 8-digit number: 90,500,000.

So, we order the numbers from least to greatest as shown.

95,000 < 905,000 < 950,000 < 9,500,000 < 90,500,000

116. There are two 7-digit numbers: 2,222,222 and 4,444,444. 4,444,444 is greater than 2,222,222.

There is only one 6-digit number: 555,555.

There are two 5-digit numbers: 11,111 and 33,333. 33,333 is greater than 11,111.

So, we order the numbers from greatest to least as shown.

4,444,444 > 2,222,222 > 555,555 > 33,333 > 11,111

117. 6 ten-thousands is 60,000.
8 thousands is 8,000.
9 hundreds is 900.
7 hundred-thousands is 700,000.
5 thousands is 5,000.

Of these numbers, 700,000 is greatest.

6 ten-thousands   8 thousands   9 hundreds

(7 hundred-thousands)   5 thousands

118. Any number of thousands is that number followed by three 0's. So, 5,500 thousands is 5,500,000.

5 millions is 5,000,000.

Any number of tens is that number followed by one 0. So, 55,555 tens is 555,550.

550 thousands is 550,000.

Any number of hundreds is that number followed by two 0's. So, 5,555 hundreds is 555,500.

Of these numbers, 5,500,000 is greatest.

(5,500 thousands)   5 millions   55,555 tens

550 thousands   5,555 hundreds

119. To make the largest four-digit number whose digits are all different, we use the four largest digits: 9, 8, 7, and 6.

To make the number as large as possible, we place the largest digit in the largest place value. So, we write 9 as the leftmost digit, followed by 8, then 7, then 6. This gives **9,876**.

120. The five even digits are 0, 2, 4, 6, and 8. To make our number as small as possible, we write the smallest digit in the leftmost place value. But, a number's leftmost digit cannot be 0. So, we write the next-smallest digit, 2, as the leftmost digit.

After 2, we write the smallest digit 0, then 4, then 6, then 8. This gives **20,468**.

121. ⬚5⬚,1 2 3 is the largest number and has ten-thousands digit 5. So, the ten-thousands digit of the smallest number, ⬚4,1 2 3, must be 5 or less.

The leftmost digit of a number cannot be 0, so we cannot fill each blank with 0. If we fill each blank with 1, 2, 3, or 4, then ⬚4,1 2 3 < ⬚,0 0 0 will not be true.

So, we fill each blank with 5 as shown.

5 4,1 2 3 < 5 5,0 0 0 < 5 5,1 2 3

122. Each digit is larger than the digit to its right. So, all of the digits must be different. To make the smallest number possible, we use the six smallest digits: 0, 1, 2, 3, 4, and 5.

We arrange these digits so that each digit is larger than the digit to its right.

543,210

123. The two smallest numbers Alex writes are in the 1,200's: 1,234 and 1,243.

The next two smallest numbers are in the 1,300's: 1,324 and 1,342.

The next two smallest numbers are in the 1,400's: 1,423 and 1,432.

So, the 5th number in Alex's list is **1,423**.

124. For Ms. Q.'s number to be as small as possible, we want the digit in the largest place value to be as small as possible. So, we try to make the leftmost digit small.

If we erase the 2, then 1 will be the leftmost digit. If we erase any digit other than the 2, then 2 will be the leftmost digit. So, we erase the 2.

2̶1089

This leaves **1,089**.

125. For the number to be as small as possible, we want the digit in the largest place value to be as small as possible. So, we try to make the leftmost digit small.

If we erase the 5, then 0 will be the leftmost digit. But, 0 cannot be the leftmost digit of a number! So, the first digit must be 5, and we try to make the second digit small.

The second digit is 0, which is the smallest digit. So, we do not erase the 0. We then try to make the third digit small.

The third digit is 6. If we erase the 6, the first three digits will be 504. If we do not erase the 6, the first three digits will be 506. So, we erase the 6.

50̶6̶48

This leaves **5,048**.

126. Since we want the digits in the largest place values to be as small as possible, we work from left to right.

The first digit is 2, which is the smallest of the given digits. So, we do not erase the 2.

2745638
  ✓

The second digit is 7. If we erase the 7, the second digit will be 4, which is less than 7. So, we erase the 7.

2̶45638
✓

Now, the second digit is 4. If we erase the 4, the second digit will be 5, which is greater than 4. So, we do not erase the 4.

2̶45638
✓ ✓

The third digit is 5. If we erase the 5, the third digit will be 6, which is greater than 5. So, we do not erase the 5.

2̶45638
✓ ✓✓

The fourth digit is 6. If we erase the 6, the fourth digit will be 3, which is less than 6. So, we erase the 6.

2̶45̶6̶38

This leaves **24,538**.

127. To make the number as small as possible, we want the leftmost digit to be as small as possible.

Among the given digits, 0 is the smallest. But, the leftmost digit of a number cannot be 0. The next-smallest digit is 1, which is already the leftmost digit. So, we do not erase the leftmost 1.

Next, we want the second digit to be as small as possible. We can make the second digit 0 by erasing all nine digits between the leftmost 1 and the 0.

1̶2̶3̶4̶5̶6̶7̶8̶9̶01112131415

This leaves **101,112,131,415**.

We could have also erased the first 9 digits to get the same result.

1̶2̶3̶4̶5̶6̶7̶8̶9̶101112131415

128. The left number has more digits than the right number. Since the left number must be smaller than the right number, we must cross out one of the left number's digits.

Only crossing out the digit below gives a true statement.

1 2 3 4̶ < 1 2 4

129. The left number has more digits than the right number. So, we must cross out one of the left number's digits.

Only crossing out the digit below gives a true statement.

6̶5 6 5 4 3 2 < 6 7 8 9 0

130. The left number has more digits than the right number. So, we must cross out one of the left number's digits.

Only crossing out the digit below gives a true statement.

2 X 2 4 2 4 < 2 3 4 5 6

131. The middle number has more digits than the right number. So, we must cross out one of the middle number's digits.

Only crossing out the digit below gives a true statement.

4 5 5 4 < 5 4 X 4 5 < 5 4 5 4

132. The only way to make the middle number smaller than the right number is by crossing out two of its digits.

So, we want to cross out two of the middle number's digits to get a result that is greater than 9,998 and less than 10,000.

9,999 is the only whole number greater than 9,998 and less than 10,000. So, we cross out the digits as shown.

9 9 9 8 < X 9 9 9 9 X < 1 0 0 0 0

133. If we cross out any of the right number's digits, then we cannot make the first two numbers smaller than the right number.

So, the right number must be 1,111. Since 4,567 and 3,457 are both greater than 1,111, we must cross out one digit in each number.

We make the left number as small as possible by crossing out the 7, leaving 456.

4 5 6 X < 3 4 5 7 < 1 1 1 1

Then, we can only make the middle number greater than 456 and less than 1,111 by crossing out the 3 as shown.

4 5 6 X < X 4 5 7 < 1 1 1 1

134. The left number is larger than the middle number. So, we must cross out one of the left number's digits.

Similarly, the middle number is larger than the right number. So, we must cross out one of the middle number's digits.

We make the left number as small as possible by crossing out the 8, giving 624.

6 X 2 4 < 4 6 8 2 < 2 4 6 8

Then, we can only make the middle number greater than 624 and less than 2,468 by crossing out the 4 as shown.

6 X 2 4 < X 6 8 2 < 2 4 6 8

135. If we cross out any of the right number's digits, then we cannot make all three of the other numbers smaller than the right number.

So, the right number must be 243. Since the other three

numbers are each greater than 243, we must cross out one digit in each number.

We make the left number as small as possible by crossing out the 4, leaving 231.

X 2 3 1 < 2 3 4 1 < 2 4 1 3 < 2 4 3

Then, we can only make the third number greater than 231 and less than 243 by crossing out the 3 as shown.

X 2 3 1 < 2 3 4 1 < 2 4 1 X < 2 4 3

Finally, we can only make the second number greater than 231 and less than 241 by crossing out the 1 as shown.

X 2 3 1 < 2 3 4 X < 2 4 1 X < 2 4 3

Close Enough 32–33

136. **35,000** is halfway between 30,000 and 40,000.

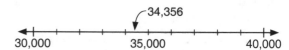

137. The number halfway between 30,000 and 40,000 is 35,000, and 34,356 is less than 35,000.

So, 34,356 is closer to **30,000** than it is to 40,000.

138. The number halfway between 600,000 and 700,000 is 650,000, and 652,476 is more than 650,000. So, 652,476 is closer to **700,000** than it is to 600,000.

139. The number halfway between 283,700 and 283,800 is 283,750, and 283,791 is more than 283,750. So, 283,791 is closer to 283,800 than it is to 283,700.

The number halfway between 280,000 and 290,000 is 285,000, and 283,791 is less than 285,000. So, 283,791 is closer to 280,000 than it is to 290,000.

The number halfway between 200,000 and 300,000 is 250,000, and 283,791 is more than 250,000. So, 283,791 is closer to 300,000 than it is to 200,000.

283,700 or (283,800)

(280,000) or 290,000

200,000 or (300,000)

140. 5,468 is between 5,460 and 5,470. Since 5,468 is closer to 5,470 than it is to 5,460, we place 7 in the first number's blank.

| 5,4**7**0 | 5, □0 | □,000 |

5,468 is between 5,400 and 5,500. Since 5,468 is closer to 5,500 than it is to 5,400, we place 5 in the second number's blank.

| 5,4 7 0 | 5,**5**00 | □,000 |

5,468 is between 5,000 and 6,000. Since 5,468 is closer to 5,000 than it is to 6,000, we place 5 in the third number's blank.

| 5,470 | 5,500 | **5,000** |

141. 87,290 is between 87,200 and 87,300. Since 87,290 is closer to 87,300 than it is to 87,200, we place 3 in the first number's blank.

| 8 7,**3** 0 0 | 8 ,0 0 0 | 0,0 0 0 |

87,290 is between 87,000 and 88,000. Since 87,290 is closer to 87,000 than it is to 88,000, we place 7 in the second number's blank.

| 8 7,3 0 0 | 8 7,0 0 0 | 0,0 0 0 |

87,290 is between 80,000 and 90,000. Since 87,290 is closer to 90,000 than it is to 80,000, we place 9 in the third number's blank.

| 8 7,3 0 0 | 8 7,0 0 0 | **9** 0,0 0 0 |

142. 45,827 is between 45,800 and 45,900. Since 45,827 is closer to 45,800 than it is to 45,900, we fill the first number's blanks as shown.

| **4 5**,8 0 0 | ,0 0 0 | 0,0 0 0 |

45,827 is between 45,000 and 46,000. Since 45,827 is closer to 46,000 than it is to 45,000, we fill the second number's blanks as shown.

| 4 5,8 0 0 | **4 6**,0 0 0 | 0,0 0 0 |

45,827 is between 40,000 and 50,000. Since 45,827 is closer to 50,000 than it is to 40,000, we fill the third number's blank as shown.

| 4 5,8 0 0 | 4 6,0 0 0 | **5** 0,0 0 0 |

143. 764,971 is between 760,000 and 770,000. Since 764,971 is closer to 760,000 than it is to 770,000, we fill the first number's blanks as shown.

| **7 6** 0,0 0 0 | ,0 0 0 | , 0 0 |

764,971 is between 764,000 and 765,000. Since 764,971 is closer to 765,000 than it is to 764,000, we fill the second number's blanks as shown.

| 7 6 0,0 0 0 | **7 6 5**,0 0 0 | , 0 0 |

764,971 is between 764,900 and 765,000. Since 764,971 is closer to 765,000 than it is to 764,900, we fill the third number's blanks as shown.

| 7 6 0,0 0 0 | 7 6 5,0 0 0 | **7 6 5,0** 0 0 |

BIG NUMBERS
Estimation   34-35

144. We consider each sum.

712,740+406,460 is not much easier to compute than the actual sum. So, it is not a good estimate. ✘

712,700+406,500 is not much easier to compute than the actual sum. So, it is not a good estimate. ✘

700,000+400,000 is easy to compute. Also, since 700,000 is close to 712,743 and 400,000 is close to 406,459, we know 700,000+400,000 is close to the exact answer. So, it is a good estimate. ✔

We circle the best estimate as shown.

712,740+406,460   712,700+406,500   ⟮700,000+400,000⟯

145. We consider each sum.

142,900+128,500 is not much easier to compute than the actual sum. So, it is not a good estimate. ✘

140,000+130,000 is easy to compute. Also, since 140,000 is close to 142,862 and 130,000 is close to 128,541, we know 140,000+130,000 is close to the exact answer. So, it is a good estimate. ✔

100,000+100,000 is easy to compute. But, 100,000 is not close to 142,862 or to 128,541. So, 100,000+100,000 is not a good estimate. ✘

We circle the best estimate as shown.

142,900+128,500   ⟮140,000+130,000⟯   100,000+100,000

146. 22,538 is close to 20,000 and 38,902 is close to 40,000. So, 22,538+38,902 is close to 20,000+40,000 = 60,000.

Since no other answer choice is close to 60,000, we circle 60,000.

20,000   40,000   ⟮60,000⟯   80,000   100,000

In fact, 22,538+38,902 = 61,440.

147. 185,057 is close to 200,000 and 476,336 is close to 500,000. So, 185,057+476,336 is close to 200,000+500,000 = 700,000.

Two of our answer choices are close to 700,000: 660,000 and 760,000.

Since 185,057 is a little less than 200,000 and 476,336 is a little less than 500,000, we know 185,057+476,336 is a little less than 200,000+500,000 = 700,000.

So, we circle 660,000.

66,000   560,000   ⟮660,000⟯   760,000   1,000,000

In fact, 185,057+476,336 = 661,393.

148. 4,097,865 is close to 4,000,000 and 1,024,196 is close to 1,000,000. So, 4,097,865+1,024,196 is close to 4,000,000+1,000,000 = 5,000,000.

There are two letters close to 5,000,000 on the number line: C and D.

Since 4,097,865 is a little more than 4,000,000 and 1,024,196 is a little more than 1,000,000, we know 4,097,865+1,024,196 is a little more than 4,000,000+1,000,000 = 5,000,000.

So, we circle the letter D.

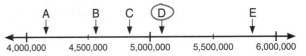

In fact, 4,097,865+1,024,196 = 5,122,061.

149. 697 is a little less than 700 and 691 is also a little less than 700. So, 697+691 is a little less than 700+700 = 1,400.

$$697+691 \, \textcircled{<} \, 1,400$$

150. 30,052 is a little more than 30,000 and 40,190 is a little more than 40,000. So, 30,052+40,190 is a little more than 30,000+40,000 = 70,000.

$$30,052+40,190 \, \textcircled{>} \, 70,000$$

151. 7,950 is a little less than 8,000 and 4,925 is a little less than 5,000. So, 7,950+4,925 is a little less than 8,000+5,000 = 13,000.

$$7,950+4,925 \, \textcircled{<} \, 13,000$$

152. 128,128 is a little less than 130,000 and 69,214 is a little less than 70,000. So, 128,128+69,214 is a little less than 130,000+70,000 = 200,000.

In other words, 200,000 is a little more than 128,128+69,214.

$$200,000 \, \textcircled{>} \, 128,128+69,214$$

153. 486 is a little less than 500 and 679 is a little less than 700. So, 486+679 is a little less than 500+700 = 1,200.

317 is a little more than 300 and 926 is a little more than 900. So, 317+926 is a little more than 300+900 = 1,200.

Since 486+679 is less than 1,200 and 317+926 is more than 1,200, we know 486+679 is less than 317+926.

$$486+679 \, \textcircled{<} \, 317+926$$

154. 5,778 is a little less than 6,000 and 8,976 is a little less than 9,000. So, 5,778+8,976 is a little less than 6,000+9,000 = 15,000.

3,123 is a little more than 3,000 and 12,430 is a little more than 12,000. So, 3,123+12,430 is a little more than 3,000+12,000 = 15,000.

Since 5,778+8,976 is less than 15,000 and 3,123+12,430 is more than 15,000, we know 5,778+8,976 is less than 3,123+12,430.

$$5,778+8,976 \, \textcircled{<} \, 3,123+12,430$$

BIG NUMBERS
Challenge Problems   36-37

155. The number halfway between 10,000 and 20,000 is 15,000. So, any number less than 15,000 is closer to 10,000 than it is to 20,000. Any number greater than 15,000 is closer to 20,000 than it is to 10,000.

So, we are looking for the largest whole number that is less than 15,000, which is **14,999**.

156. Thirty-three thousand thirty-three is 33,033. Forty-four million forty-four is 44,000,044.

So, we have 33,033+44,000,044 = **44,033,077**.

157. The largest number given is 3,333,333. The smallest number given is 99,999.

So, we are looking for 3,333,333 − 99,999. To subtract 99,999, we can subtract 100,000 then add 1 back.

3,333,333 − 100,000 = 3,233,333, and 3,233,333 + 1 = **3,233,334**.

158. The three largest 5-digit numbers are 99,999, 99,998, and 99,997.

99,999 is 1 less than 100,000.
99,998 is 2 less than 100,000.
99,997 is 3 less than 100,000.

So, 99,999+99,998+99,997 is 1+2+3 = 6 less than 100,000+100,000+100,000 = 300,000.

So, 99,999+99,998+99,997 = 300,000−6 = **299,994**.

159. When Alex switches the thousands and the ones digit of 6,67<u>7</u>, he gets <u>7</u>,676.

To find how much greater 7,676 is than 6,677, we consider their difference on the number line by counting up from 6,677. Adding 1,000 to 6,677 gives 7,677. Taking 1 away from 7,677 gives 7,676.

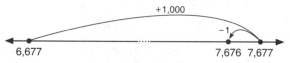

So, to go from 6,677 to 7,676, we add 1,000 then take away 1. This is the same as adding 999.

So, 7,676 is **999** greater than 6,677.

— *or* —

Switching the thousands digit from 6 to 7 adds 1,000 to the number.

Switching the ones digit from 7 to 6 takes 1 away from the number.

Adding 1,000 then taking 1 away is the same as adding 999. So, Alex's result is **999** greater than 6,677.

160. We use an estimate for each number.

10,927 is close to 11,000.
17,357 is close to 17,000.
20,847 is close to 21,000.
24,437 is close to 24,000.
29,187 is close to 29,000.

We're looking for two numbers that sum to 38,204. Among our estimates, only 17,000+21,000 = 38,000 is close to 38,204.

So, we circle 17,357 and 20,847.

10,927 $\boxed{17,357}$ $\boxed{20,847}$ 24,437   29,187

161. The closest number we can make to 5,000 is either a little more than 5,000 or a little less than 5,000.

The smallest number we can make that is a little more than 5,000 is 5,456.

The largest number we can make that is a little less than 5,000 is 4,655.

To find which of 5,456 and 4,655 is closest to 5,000, we compute the difference between each number and 5,000.

5,456 − 5,000 = 456.
5,000 − 4,655 = 345.

Since 5,000 and 4,655 have the smallest difference, **4,655** is the closest number we can make to 5,000.

162. 10 thousands is 10,000.
10 hundreds is 1,000.
10 tens is 100.

So, 10 thousands, 10 hundreds, and 10 tens is
10,000 + 1,000 + 100 = **11,100**.

163. The smallest 7-digit number is 1,000,000. But, these digits add up to 1, not 20. So, we must make some of the digits larger.

To keep our number as small as possible, we increase the place values farthest to the right first, starting with the ones place.

If the ones digit is 9, we have 1,000,009. These digits add up to 1 + 9 = 10, not 20. So, we must increase the tens digit as well.

If the tens digit is also 9, we have 1,000,099. These digits add up to 1 + 9 + 9 = 19, which is only 1 less than 20. So, we only need to increase the value of a digit by 1.

Increasing the hundreds digit by 1 gives 1,000,199. These digits add up to 1 + 1 + 9 + 9 = 20.

Since the leftmost digits are as small as possible, our 7-digit number is as small as possible. So, **1,000,199** is the smallest 7-digit number whose digits add up to 20.

Step-By-Step   39

1. We use the algorithm to continue the list.

10 is even, so half of 10 is 5.
5 is odd, so three copies of 5 plus 1 is $5+5+5+1 = 16$.
16 is even, so half of 16 is 8.
8 is even, so half of 8 is 4.
4 is even, so half of 4 is 2.
2 is even, so half of 2 is 1.
1 is odd, so three copies of 1 plus 1 is $1+1+1+1 = 4$.

We fill the blanks in the list as shown.

  6 , 3 , 10 , **5** , **16** , **8** , **4** , **2** , **1** , **4**

2. We use the algorithm to continue the list.

22 is even, so half of 22 is 11.
11 is odd, so three copies of 11 plus 1 is $11+11+11+1 = 34$.
34 is even, so half of 34 is 17.
17 is odd, so three copies of 17 plus 1 is $17+17+17+1 = 52$.
52 is even, so half of 52 is 26.
26 is even, so half of 26 is 13.
13 is odd, so three copies of 13 plus 1 is $13+13+13+1 = 40$.
40 is even, so half of 40 is 20.

We fill the blanks in the list as shown.

  7 , 22 , **11** , **34** , **17** , **52** , **26** , **13** , **40** , **20** .

3. We use the algorithm to continue the list.

4 is even, so half of 4 is 2.
2 is even, so half of 2 is 1.
1 is odd, so three copies of 1 plus 1 is $1+1+1+1 = 4$.
4 is even, so half of 4 is 2.
2 is even, so half of 2 is 1.
1 is odd, so three copies of 1 plus 1 is $1+1+1+1 = 4$.

We notice the list keeps repeating! After 4 we get 2. After 2 we get 1. After 1 we get 4. So, the list is 4, 2, 1, repeated over and over.

We fill the blanks in the list as shown.

  4 , **2** , **1** , **4** , **2** , **1** , **4** , **2** , **1** , **4**

Stacking   40-45

4. Adding the ones gives $7+6 = 13$.

Adding the tens gives $30+80 = 110$.

All together, we have $13+110 = 123$.

$$\begin{array}{r} 3\ 7 \\ +\ 8\ 6 \\ \hline 1\boxed{3} \\ +\boxed{1}\boxed{1}\boxed{0} \\ \hline \boxed{1}\boxed{2}\boxed{3} \end{array}$$

5. Adding the ones gives $4+9 = 13$.

Adding the tens gives $50+30 = 80$.

All together, we have $13+80 = 93$.

$$\begin{array}{r} 5\ 4 \\ +\ 3\ 9 \\ \hline 1\ 3 \\ +\ 8\ 0 \\ \hline 9\ 3 \end{array}$$

6. Adding the ones gives $9+3 = 12$.

Adding the tens gives $80+50 = 130$.

All together, we have $12+130 = 142$.

$$\begin{array}{r} 8\ 9 \\ +\ 5\ 3 \\ \hline 1\boxed{2} \\ +\boxed{1}\boxed{3}\boxed{0} \\ \hline \boxed{1}\boxed{4}\boxed{2} \end{array}$$

7. Adding the ones gives $7+6 = 13$ ones, which is 1 ten and 3 ones.

We put the 3 at the bottom of the ones column, and the 1 at the top of the tens column.

$$\begin{array}{r} \boxed{1} \\ 3\ 7 \\ +\ \ 8\ 6 \\ \hline \square\square\boxed{3} \end{array}$$

Adding the tens gives $1+3+8 = 12$ tens, which is 1 hundred and 2 tens.

We put the 2 at the bottom of the tens column. Since there are no more hundreds, we put the 1 at the bottom of the hundreds column.

$$\begin{array}{r} \boxed{1} \\ 3\ 7 \\ +\ \ 8\ 6 \\ \hline \boxed{1}\boxed{2}\boxed{3} \end{array}$$

8. Adding the ones gives $4+9 = 13$ ones, which is 1 ten and 3 ones.

We put the 3 at the bottom of the ones column, and the 1 at the top of the tens column.

$$\begin{array}{r} \boxed{1} \\ 5\ 4 \\ +\ \ 3\ 9 \\ \hline \square\boxed{3} \end{array}$$

Adding the tens gives $1+5+3 = 9$ tens.

We put the 9 at the bottom of the tens column.

$$\begin{array}{r} \boxed{1} \\ 5\ 4 \\ +\ \ 3\ 9 \\ \hline \boxed{9}\boxed{3} \end{array}$$

9. Adding the ones gives $9+3 = 12$ ones, which is 1 ten and 2 ones.

We put the 2 at the bottom of the ones column, and the 1 at the top of the tens column.

$$\begin{array}{r} \boxed{1} \\ 8\ 9 \\ +\ \ 5\ 3 \\ \hline \square\square\boxed{2} \end{array}$$

Adding the tens gives $1+8+5 = 14$ tens, which is 1 hundred and 4 tens.

We put the 4 at the bottom of the tens column. Since there are no more hundreds, we put the 1 at the bottom of the hundreds column.

$$\begin{array}{r} \boxed{1} \\ 8\ 9 \\ +\ \ 5\ 3 \\ \hline \boxed{1}\boxed{4}\boxed{2} \end{array}$$

10.

Step 1:	Step 2:	Final:
We add the ones.	We add the tens.	We add the hundreds.

Step 1:
$$\begin{array}{r} \square\boxed{1} \\ 2\ 4\ 6 \\ +\ 5\ 7\ 9 \\ \hline \square\square\boxed{5} \end{array}$$

Step 2:
$$\begin{array}{r} \boxed{1}\boxed{1} \\ 2\ 4\ 6 \\ +\ 5\ 7\ 9 \\ \hline \square\boxed{2}\boxed{5} \end{array}$$

Final:
$$\begin{array}{r} \boxed{1}\boxed{1} \\ 2\ 4\ 6 \\ +\ 5\ 7\ 9 \\ \hline \boxed{8}\boxed{2}\boxed{5} \end{array}$$

11. Step 1:
We add the ones.

☐1
 1 5 8
+ 3 4 7
☐☐5

Step 2:
We add the tens.

11
 1 5 8
+ 3 4 7
☐05

Final:
We add the hundreds.

11
 1 5 8
+ 3 4 7
505

12. Step 1:
We add the ones.

1
 3 5 8
+ 9 3 4
☐,☐☐2

Step 2:
We add the tens.

1
 3 5 8
+ 9 3 4
☐,☐92

Final:
We add the hundreds.

1
 3 5 8
+ 9 3 4
1,292

13. Step 1:

☐
 4 7 3
+ 6 8 4
☐,☐☐7

Step 2:

1
 4 7 3
+ 6 8 4
☐,☐57

Final:

1
 4 7 3
+ 6 8 4
1,157

14. Step 1:

☐1
 2 5 6
+ 4 5 9
☐☐5

Step 2:

11
 2 5 6
+ 4 5 9
☐15

Final:

11
 2 5 6
+ 4 5 9
715

15. Step 1:

☐1
 7 8 5
+ 8 1 6
☐,☐☐1

Step 2:

11
 7 8 5
+ 8 1 6
☐,☐01

Final:

11
 7 8 5
+ 8 1 6
1,601

16. Step 1:

☐
 8 7 6
+ 5 4 3
☐,☐☐9

Step 2:

1
 8 7 6
+ 5 4 3
☐,☐19

Final:

1
 8 7 6
+ 5 4 3
1,419

17. Step 1:

1
 2 2 7
+ 4 1 4
☐☐1

Step 2:

1
 2 2 7
+ 4 1 4
☐41

Final:

1
 2 2 7
+ 4 1 4
641

18. Step 1:

☐1
 2 4 6
+ 3 6 9
☐☐5

Step 2:

11
 2 4 6
+ 3 6 9
☐15

Final:

11
 2 4 6
+ 3 6 9
615

19. Step 1:

☐1
 4 4 4
+ 7 7 7
☐,☐☐1

Step 2:

11
 4 4 4
+ 7 7 7
☐,☐21

Final:

11
 4 4 4
+ 7 7 7
1,221

20. Step 1:

☐1
 7 6 5
+ 1 4 7
☐☐2

Step 2:

11
 7 6 5
+ 1 4 7
☐12

Final:

11
 7 6 5
+ 1 4 7
912

21. Step 1:

1
 8 3 3
+ 3 3 8
☐,☐☐1

Step 2:

1
 8 3 3
+ 3 3 8
☐,☐71

Final:

1
 8 3 3
+ 3 3 8
1,171

22. Step 1:

☐ 1
 3 , 6 0 9
+   8 7 4
☐,☐☐3

Step 2:

☐ 1
 3 , 6 0 9
+   8 7 4
☐,☐83

Step 3:

1 1
 3 , 6 0 9
+   8 7 4
☐,483

Final:

1 1
 3 , 6 0 9
+   8 7 4
4,483

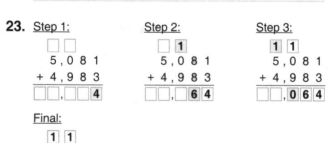

23. Step 1:

☐☐
 5 , 0 8 1
+ 4 , 9 8 3
☐☐,☐☐4

Step 2:

☐1
 5 , 0 8 1
+ 4 , 9 8 3
☐☐,☐64

Step 3:

11
 5 , 0 8 1
+ 4 , 9 8 3
☐☐,064

Final:

11
 5 , 0 8 1
+ 4 , 9 8 3
10,064

24. We fill the blanks as shown below.

$$\begin{array}{r} \boxed{1}\,\boxed{1}\,\boxed{1}\ \ \ \ \ \boxed{1} \\ 8,675,309 \\ +\ 8,675,309 \\ \hline \boxed{1}\boxed{7},\boxed{3}\boxed{5}\boxed{0},\boxed{6}\boxed{1}\boxed{8} \end{array}$$

25. We fill the blanks as shown below.

$$\begin{array}{r} \boxed{1}\boxed{1}\,\boxed{1}\ \boxed{1}\boxed{1}\,\boxed{1}\ \boxed{1}\boxed{1} \\ 123,456,789 \\ +\ 987,654,321 \\ \hline \boxed{1},\boxed{1}\boxed{1}\boxed{1},\boxed{1}\boxed{1}\boxed{1},\boxed{1}\boxed{1}\boxed{0} \end{array}$$

26. We fill the blanks as shown below.

$$\begin{array}{r} \boxed{1}\boxed{1}\ \ \ \ \boxed{1}\boxed{1}\ \boxed{1}\boxed{1}\boxed{1}\ \ \ \boxed{1} \\ 65,656,565,656 \\ +\ \ \ 7,437,437,437 \\ \hline \boxed{7}\boxed{3},\boxed{0}\boxed{9}\boxed{4},\boxed{0}\boxed{0}\boxed{3},\boxed{0}\boxed{9}\boxed{3} \end{array}$$

27. First we stack the digits. Since 1,659 has the most digits, we place it on top. Then, we add by place value.

Step 1:
$$\begin{array}{r} \boxed{1},\boxed{6}\boxed{5}\boxed{9} \\ +\ \ \ \boxed{2}\boxed{8}\boxed{2} \\ \hline \boxed{},\boxed{}\boxed{}\boxed{} \end{array}$$

Step 2:
$$\begin{array}{r} 1 \\ 1,659 \\ +\ \ \ 282 \\ \hline \boxed{},\boxed{}\boxed{}\boxed{1} \end{array}$$

Step 3:
$$\begin{array}{r} 1\ \ 1 \\ 1,659 \\ +\ \ \ 282 \\ \hline \boxed{},\boxed{}\boxed{4}\boxed{1} \end{array}$$

Step 4:
$$\begin{array}{r} 1\ \ 1 \\ 1,659 \\ +\ \ \ 282 \\ \hline \boxed{},\boxed{9}\boxed{4}\boxed{1} \end{array}$$

Final:
$$\begin{array}{r} 1\ \ 1 \\ 1,659 \\ +\ \ \ 282 \\ \hline \boxed{1},\boxed{9}\boxed{4}\boxed{1} \end{array}$$

So, 1,659+282 = **1,941**.

28. First we stack the digits. Since 1,297 has the most digits, we place it on top. Then, we add by place value.

Step 1:
$$\begin{array}{r} \boxed{1},\boxed{2}\boxed{9}\boxed{7} \\ +\ \ \ \boxed{7}\boxed{4}\boxed{6} \\ \hline \boxed{},\boxed{}\boxed{}\boxed{} \end{array}$$

Step 2:
$$\begin{array}{r} 1 \\ 1,297 \\ +\ \ \ 746 \\ \hline \boxed{},\boxed{}\boxed{}\boxed{3} \end{array}$$

Step 3:
$$\begin{array}{r} 1\ \ 1 \\ 1,297 \\ +\ \ \ 746 \\ \hline \boxed{},\boxed{}\boxed{4}\boxed{3} \end{array}$$

Step 4:
$$\begin{array}{r} 1\ 1\ 1 \\ 1,297 \\ +\ \ \ 746 \\ \hline \boxed{},\boxed{0}\boxed{4}\boxed{3} \end{array}$$

Final:
$$\begin{array}{r} 1\ 1\ 1 \\ 1,297 \\ +\ \ \ 746 \\ \hline \boxed{2},\boxed{0}\boxed{4}\boxed{3} \end{array}$$

So, 746+1,297 = **2,043**.

29. First we stack the digits. Then, we add by place value.

Step 1:
$$\begin{array}{r} 531 \\ +\ 284 \\ \hline \end{array}$$

Step 2:
$$\begin{array}{r} 531 \\ +\ 284 \\ \hline 5 \end{array}$$

Step 3:
$$\begin{array}{r} 1 \\ 531 \\ +\ 284 \\ \hline 15 \end{array}$$

Final:
$$\begin{array}{r} 1 \\ 531 \\ +\ 284 \\ \hline 815 \end{array}$$

So, 531+284 = **815**.

30. First we stack the digits. Then, we add by place value.

Step 1:
$$\begin{array}{r} 538 \\ +\ 765 \\ \hline \end{array}$$

Step 2:
$$\begin{array}{r} 1 \\ 538 \\ +\ 765 \\ \hline 3 \end{array}$$

Step 3:
$$\begin{array}{r} 1\ 1 \\ 538 \\ +\ 765 \\ \hline 03 \end{array}$$

Final:
$$\begin{array}{r} 1\ 1 \\ 538 \\ +\ 765 \\ \hline 1,303 \end{array}$$

So, 538+765 = **1,303**.

31. First we stack the digits. Since 8,376 has the most digits, we place it on top. Then, we add by place value.

Step 1:
$$\begin{array}{r} 8,376 \\ +\ \ \ 247 \\ \hline \end{array}$$

Step 2:
$$\begin{array}{r} 1 \\ 8,376 \\ +\ \ \ 247 \\ \hline 3 \end{array}$$

Step 3:
$$\begin{array}{r} 1\ 1 \\ 8,376 \\ +\ \ \ 247 \\ \hline 23 \end{array}$$

Step 4:
$$\begin{array}{r} 1\ 1 \\ 8,376 \\ +\ \ \ 247 \\ \hline 623 \end{array}$$

Final:
$$\begin{array}{r} 1\ 1 \\ 8,376 \\ +\ \ \ 247 \\ \hline 8,623 \end{array}$$

So, 8,376+247 = **8,623**.

32. First we stack the digits. Since 1,934 has the most digits, we place it on top. Then, we add by place value.

Step 1:
$$\begin{array}{r} 1,934 \\ +\ \ \ 527 \\ \hline \end{array}$$

Step 2:
$$\begin{array}{r} 1 \\ 1,934 \\ +\ \ \ 527 \\ \hline 1 \end{array}$$

Step 3:
$$\begin{array}{r} 1 \\ 1,934 \\ +\ \ \ 527 \\ \hline 61 \end{array}$$

Step 4:
$$\begin{array}{r} 1\ \ 1 \\ 1,934 \\ +\ \ \ 527 \\ \hline 461 \end{array}$$

Final:
$$\begin{array}{r} 1\ \ 1 \\ 1,934 \\ +\ \ \ 527 \\ \hline 2,461 \end{array}$$

So, 527+1,934 = **2,461**.

33. In the ones column, 2+5 = $\boxed{7}$.

$$\begin{array}{r} \square\,7\,2 \\ +\ 3\ \square\ 5 \\ \hline 9\ 8\ \boxed{7} \end{array}$$

In the tens column, 7+$\boxed{1}$ = 8.

$$\begin{array}{r} \square\,7\,2 \\ +\ 3\ \boxed{1}\ 5 \\ \hline 9\ 8\ \boxed{7} \end{array}$$

In the hundreds column, $\boxed{6}$+3 = 9.

$$\begin{array}{r} \boxed{6}\,7\,2 \\ +\ 3\ \boxed{1}\ 5 \\ \hline 9\ 8\ \boxed{7} \end{array}$$

34. In the ones column, 6+9 = 1$\boxed{5}$.
We place a 1 above the tens column.

$$\begin{array}{r} 1 \\ 5\ \square\ 6 \\ +\ \square\ 4\ 9 \\ \hline 7\ 8\ \boxed{5} \end{array}$$

In the tens column, 1+$\boxed{3}$+4 = 8.

$$\begin{array}{r} 1 \\ 5\ \boxed{3}\ 6 \\ +\ \square\ 4\ 9 \\ \hline 7\ 8\ \boxed{5} \end{array}$$

In the hundreds column, 5+$\boxed{2}$ = 7.

$$\begin{array}{r} 1 \\ 5\ \boxed{3}\ 6 \\ +\ \boxed{2}\ 4\ 9 \\ \hline 7\ 8\ \boxed{5} \end{array}$$

35. Step 1: $\boxed{3}$+4 = 7.

$$\begin{array}{r} 7\ 0\ \boxed{3} \\ +\ \square\ 3\ 4 \\ \hline 1,5\ \square\ 7 \end{array}$$

Step 2: 0+3 = $\boxed{3}$.

$$\begin{array}{r} 7\ 0\ \boxed{3} \\ +\ \square\ 3\ 4 \\ \hline 1,5\ \boxed{3}\ 7 \end{array}$$

Final: 7+$\boxed{8}$ = 15.

$$\begin{array}{r} 7\ 0\ \boxed{3} \\ +\ \boxed{8}\ 3\ 4 \\ \hline 1,5\ \boxed{3}\ 7 \end{array}$$

36. Step 1: $\boxed{5}$+4 = 9.

$$\begin{array}{r} 9\ 2\ \boxed{5} \\ +\ \square\ 2\ 4 \\ \hline \square,5\ \square\ 9 \end{array}$$

Step 2: 2+2 = $\boxed{4}$.

$$\begin{array}{r} 9\ 2\ \boxed{5} \\ +\ \square\ 2\ 4 \\ \hline \square,5\ \boxed{4}\ 9 \end{array}$$

Final: 9+6 = 1$\boxed{5}$.

$$\begin{array}{r} 9\ 2\ \boxed{5} \\ +\ \boxed{6}\ 2\ 4 \\ \hline \boxed{1},5\ \boxed{4}\ 9 \end{array}$$

37. Step 1: 6+4 = 1$\boxed{0}$.

$$\begin{array}{r} 1 \\ \square\ 4\ 6 \\ +\ 5\ \square\ 4 \\ \hline 8\ 4\ \boxed{0} \end{array}$$

Step 2: 1+4+$\boxed{9}$ = 14.

$$\begin{array}{r} 1\ 1 \\ \square\ 4\ 6 \\ +\ 5\ \boxed{9}\ 4 \\ \hline 8\ 4\ \boxed{0} \end{array}$$

Final: 1+$\boxed{2}$+5 = 8.

$$\begin{array}{r} 1\ 1 \\ \boxed{2}\ 4\ 6 \\ +\ 5\ \boxed{9}\ 4 \\ \hline 8\ 4\ \boxed{0} \end{array}$$

38. Step 1: $\boxed{5}$+8 = 13.

$$\begin{array}{r} 1 \\ 7\ 0\ \boxed{5} \\ +\ \square\ \square\ 8 \\ \hline 1,0\ 5\ 3 \end{array}$$

Step 2: 1+0+$\boxed{4}$ = 5.

$$\begin{array}{r} 1 \\ 7\ 0\ \boxed{5} \\ +\ \square\ \boxed{4}\ 8 \\ \hline 1,0\ 5\ 3 \end{array}$$

Final: 7+$\boxed{3}$ = 10.

$$\begin{array}{r} 1 \\ 7\ 0\ \boxed{5} \\ +\ \boxed{3}\ \boxed{4}\ 8 \\ \hline 1,0\ 5\ 3 \end{array}$$

39. Step 1: $\boxed{5}$+4 = 9.

$$\begin{array}{r} 7,\ \square\ 0\ \boxed{5} \\ +\ \square,8\ 6\ 4 \\ \hline 1\ 4,2\ \square\ 9 \end{array}$$

Step 2: 0+6 = $\boxed{6}$.

$$\begin{array}{r} 7,\ \square\ 0\ \boxed{5} \\ +\ \square,8\ 6\ 4 \\ \hline 1\ 4,2\ \boxed{6}\ 9 \end{array}$$

Step 3: $\boxed{4}$+8 = 12.

$$\begin{array}{r} 1 \\ 7,\boxed{4}\ 0\ \boxed{5} \\ +\ \square,8\ 6\ 4 \\ \hline 1\ 4,2\ 6\ 9 \end{array}$$

Final: 1+7+$\boxed{6}$ = 14.

$$\begin{array}{r} 1 \\ 7,\boxed{4}\ 0\ \boxed{5} \\ +\ \boxed{6},8\ 6\ 4 \\ \hline 1\ 4,2\ 6\ 9 \end{array}$$

40. Step 1: $\boxed{5}$+8 = 13.

$$\begin{array}{r} 1 \\ 8,6\ 4\ \boxed{5} \\ +\ \square,\square\ \square\ 8 \\ \hline 1\ 1,2\ 6\ 3 \end{array}$$

Step 2: 1+4+$\boxed{1}$ = 6.

$$\begin{array}{r} 1 \\ 8,6\ 4\ \boxed{5} \\ +\ \square,\square\ \boxed{1}\ 8 \\ \hline 1\ 1,2\ 6\ 3 \end{array}$$

Step 3: 6+$\boxed{6}$ = 12.

$$\begin{array}{r} 1\quad\ 1 \\ 8,6\ 4\ \boxed{5} \\ +\ \square,\boxed{6}\ \boxed{1}\ 8 \\ \hline 1\ 1,2\ 6\ 3 \end{array}$$

Final: 1+8+$\boxed{2}$ = 11.

$$\begin{array}{r} 1\quad\ 1 \\ 8,6\ 4\ \boxed{5} \\ +\ \boxed{2},\boxed{6}\ \boxed{1}\ 8 \\ \hline 1\ 1,2\ 6\ 3 \end{array}$$

41. Step 1: $\boxed{5}$+8 = 13.

$$\begin{array}{r} 1 \\ 7,\square\ \square\ \boxed{5} \\ +\ \square,8\ 8\ 8 \\ \hline 1\ 0,3\ 3\ 3 \end{array}$$

Step 2: 1+$\boxed{4}$+8 = 13.

$$\begin{array}{r} 1\ 1 \\ 7,\square\ \boxed{4}\ \boxed{5} \\ +\ \square,8\ 8\ 8 \\ \hline 1\ 0,3\ 3\ 3 \end{array}$$

Step 3: 1+$\boxed{4}$+8 = 13.

$$\begin{array}{r} 1\ 1\ 1 \\ 7,\boxed{4}\ \boxed{4}\ \boxed{5} \\ +\ \square,8\ 8\ 8 \\ \hline 1\ 0,3\ 3\ 3 \end{array}$$

Final: 1+7+$\boxed{2}$ = 10.

$$\begin{array}{r} 1\ 1\ 1 \\ 7,\boxed{4}\ \boxed{4}\ \boxed{5} \\ +\ \boxed{2},8\ 8\ 8 \\ \hline 1\ 0,3\ 3\ 3 \end{array}$$

42. Step 1:

[4]+9 = 13.

```
      1
  4 , 2 [ ][4]
+ [ ],[ ] 1  9
--------------
  6 , 1  3  3
```

Step 2:

1+[1]+1 = 3.

```
      1
  4 , 2 [1][4]
+ [ ],[ ] 1  9
--------------
  6 , 1  3  3
```

Step 3:

2+[9] = 11.

```
  1    1
  4 , 2 [1][4]
+ [ ],[9] 1  9
--------------
  6 , 1  3  3
```

Final:

1+4+[1] = 6.

```
  1    1
  4 , 2 [1][4]
+ [1],[9] 1  9
--------------
  6 , 1  3  3
```

43. Step 1:

[6]+8 = 14.

```
          1
    1 ,[ ] 3 [6]
+   [ ], 8  1  8
----------------
  1 [ ], 8 [ ] 4
```

Step 2:

1+3+1 = [5].

```
          1
    1 ,[ ] 3 [6]
+   [ ], 8  1  8
----------------
  1 [ ], 8 [5] 4
```

Step 3:

[0]+8 = 8.

```
          1
    1 ,[0] 3 [6]
+   [ ], 8  1  8
----------------
  1 [ ], 8 [5] 4
```

Final:

1+[9] = 1[0].

```
          1
    1 ,[0] 3 [6]
+   [9], 8  1  8
----------------
  1 [0], 8 [5] 4
```

44. Step 1:

8+[8] = 16.

```
           1
     [ ],[ ] 2  8
+     5 , 4 [ ][8]
------------------
  [ ] 1 , 1  7  6
```

Step 2:

1+2+[4] = 7.

```
           1
     [ ],[ ] 2  8
+     5 , 4 [4][8]
------------------
  [ ] 1 , 1  7  6
```

Step 3:

[7]+4 = 11.

```
       1   1
     [ ],[7] 2  8
+     5 , 4 [4][8]
------------------
  [ ] 1 , 1  7  6
```

Final:

1+[5]+5 = [1]1.

```
       1   1
     [5],[7] 2  8
+     5 , 4 [4][8]
------------------
  [1] 1 , 1  7  6
```

45. Step 1:

[9]+4 = 13.

```
              1
   [ ],[ ][ ][9]
+        6  5  4
----------------
[ ][ ], 5  4  3
```

Step 2:

1+[8]+5 = 14.

```
           1  1
   [ ],[ ][8][9]
+        6  5  4
----------------
[ ][ ], 5  4  3
```

Step 3:

1+[8]+6 = 15.

```
        1  1  1
   [ ],[8][8][9]
+        6  5  4
----------------
[ ][ ], 5  4  3
```

Final:

In the thousands column, only 1+[9] = [1][0] gives a sum that is 10 or more.

```
        1  1  1
   [9],[8][8][9]
+        6  5  4
----------------
[1][0], 5  4  3
```

46. Step 1:

In the tens column, only 1+0+0 = 1. So, the tens column needs another ten.

```
    7 ,[ ] 0 [ ]
+   [ ], 8  0  1
----------------
  1 7 , 1  1 [ ]
```

In the ones column, only [9]+1 = 1[0] gives a sum that is 10 or more.

```
             1
    7 ,[ ] 0 [9]
+   [ ], 8  0  1
----------------
  1 7 , 1  1 [0]
```

Step 2:

[3]+8 = 11.

```
    1      1
    7 ,[3] 0 [9]
+   [ ], 8  0  1
----------------
  1 7 , 1  1 [0]
```

Final:

1+7+[9] = 17.

```
    1      1
    7 ,[3] 0 [9]
+   [9], 8  0  1
----------------
  1 7 , 1  1 [0]
```

ALGORITHMS
More Than Two 48–49

47. Step 1:

We add the ones.

```
       8  4
       3  3
+      7  1
-----------
 [ ][ ][8]
```

Final:

We add the tens.

```
       8  4
       3  3
+      7  1
-----------
 [1][8][8]
```

48. Step 1:

We add the ones.

```
       2
    3  4  9
    2  5  8
+      6  7
-----------
 [ ][ ][4]
```

Step 2:

We add the tens.

```
    1  2
    3  4  9
    2  5  8
+      6  7
-----------
 [ ][7][4]
```

Final:

We add the hundreds.

```
    1  2
    3  4  9
    2  5  8
+      6  7
-----------
 [6][7][4]
```

49.

Step 1:
We add
the ones.

```
    1
  8 4 1
  3 5 8
+ 9 3 4
──────
□,□□3
```

Step 2:
We add
the tens.

```
  1 1
  8 4 1
  3 5 8
+ 9 3 4
──────
□,□33
```

Final:
We add the
hundreds.

```
  1 1
  8 4 1
  3 5 8
+ 9 3 4
──────
2,133
```

50.

Step 1:
```
    1
  6 5 1
  5 2 5
+   7 4
──────
□,□□0
```

Step 2:
```
  1 1
  6 5 1
  5 2 5
+   7 4
──────
□,□50
```

Final:
```
  1 1
  6 5 1
  5 2 5
+   7 4
──────
1,250
```

51.

Step 1:
```
    1
  3 8 1
  5 6 9
+ 7 5 2
──────
□,□□2
```

Step 2:
```
  2 1
  3 8 1
  5 6 9
+ 7 5 2
──────
□,□02
```

Final:
```
  2 1
  3 8 1
  5 6 9
+ 7 5 2
──────
1,702
```

52.

Step 1:
```
    2
  4 5 6
  7 7 7
+ 1 4 9
──────
□,□□2
```

Step 2:
```
  1 2
  4 5 6
  7 7 7
+ 1 4 9
──────
□,□82
```

Final:
```
  1 2
  4 5 6
  7 7 7
+ 1 4 9
──────
1,382
```

53. First we stack the digits. Then, we add by place value.

Step 1:
```
    5 9
    8 2
    1 6
+   3 7
──────
  □□□
```

Step 2:
```
    2
    5 9
    8 2
    1 6
+   3 7
──────
   □4
```

Final:
```
    2
    5 9
    8 2
    1 6
+   3 7
──────
  1 9 4
```

So, 59+82+16+37 = **194**.

54. First we stack the digits. Since 5,389 has the most digits, we place it on top. 742 has the second most digits, so we place it second. Then, we add by place value.

Step 1:
```
  5,3 8 9
    7 4 2
+     6 7
────────
  □,□□□
```

Step 2:
```
      1
  5,3 8 9
    7 4 2
+     6 7
────────
  □,□□8
```

Step 3:
```
    1 1
  5,3 8 9
    7 4 2
+     6 7
────────
  □,□98
```

Step 4:
```
  1 1 1
  5,3 8 9
    7 4 2
+     6 7
────────
  □,198
```

Final:
```
  1 1 1
  5,3 8 9
    7 4 2
+     6 7
────────
  6,198
```

So, 742+5,389+67 = **6,198**.

55. First we stack the digits. Then, we add.

Step 1:
```
  5 3 1
  2 8 4
+ 6 7 2
──────
      7
```

Step 2:
```
    1
  5 3 1
  2 8 4
+ 6 7 2
──────
    8 7
```

Final:
```
    1
  5 3 1
  2 8 4
+ 6 7 2
──────
1,4 8 7
```

So, 531+284+672 = **1,487**.

56. First we stack the digits. Since 1,049 has the most digits, we place it on top. Then, we add by place value.

Step 1:
```
      2
  1,0 4 9
    8 0 7
+   4 2 5
────────
        1
```

Step 2:
```
      2
  1,0 4 9
    8 0 7
+   4 2 5
────────
      8 1
```

Step 3:
```
  1   2
  1,0 4 9
    8 0 7
+   4 2 5
────────
    2 8 1
```

Final:
```
  1   2
  1,0 4 9
    8 0 7
+   4 2 5
────────
  2,2 8 1
```

So, 807+1,049+425 = **2,281**.

Digit Fill   50-51

57. The sum is between ☐1+9 = 10 and ☐6+9 = 15, so the tens digit of the sum is 1.

```
  □
+ 9
───
1 □
```

The remaining digits are 5 and 6. There is only one way to place them in the empty boxes to give a correct sum.

```
  6
+ 9
───
1 5
```

58. In the ones column, either 4+☐0 = 4 or 4+☐6 = 10. So, we check the tens column instead.

In the tens column, we can only get a true statement if we place a 4 in the empty box and add 1 ten from the ones column. So, 1+☐4+5 = 10.

```
  1 1
  1 4 4
+   5 □
─────
  2 0 □
```

Then, in the ones column, only 4+☐6 = 1☐0 gives a true statement.

```
  1 1
  1 4 4
+   5 6
─────
  2 0 0
```

59. In the ones column, only ☐8+3 = 1☐1 gives a true statement.

```
    1
  3 4 8
+ 5 5 3
─────
  □□1
```

In the tens column, $1+4+5=1\boxed{0}$.

$$\begin{array}{r} {\scriptstyle 1\ \ 1}\\ 3\ 4\ \boxed{8}\\ +\ 5\ 5\ 3\\ \hline \boxed{\ }\ 0\ 1 \end{array}$$

In the hundreds column, $1+3+5=\boxed{9}$.

$$\begin{array}{r} {\scriptstyle 1\ \ 1}\\ 3\ 4\ \boxed{8}\\ +\ 5\ 5\ 3\\ \hline \boxed{9}\ 0\ \boxed{1} \end{array}$$

60. In the ones column, only $\boxed{8}+7=15$ gives a sum that ends in 5.

$$\begin{array}{r} {\scriptstyle 1}\\ 1\ 4\ \boxed{8}\\ +\ 3\ \boxed{\ }\ 7\\ \hline \boxed{\ }\ \boxed{\ }\ 5 \end{array}$$

In the tens column, either $1+4+\boxed{4}=\boxed{9}$ or $1+4+\boxed{9}=1\boxed{4}$. Since we must place the 4 and the 9 in the tens column, we can only place the 5 in the hundreds column.

$$\begin{array}{r} {\scriptstyle 1}\\ 1\ 4\ \boxed{8}\\ +\ 3\ \boxed{\ }\ 7\\ \hline \boxed{5}\ \boxed{\ }\ 5 \end{array}$$

In the hundreds column, only $1+1+3$ gives a sum of 5. This means the sum of the tens column must be 10 or more. So, in the tens column, $1+4+\boxed{9}=1\boxed{4}$.

$$\begin{array}{r} {\scriptstyle 1\ \ 1}\\ 1\ 4\ \boxed{8}\\ +\ 3\ \boxed{9}\ 7\\ \hline \boxed{5}\ 4\ \boxed{5} \end{array}$$

61. In the hundreds column, only $8+\boxed{1}=\boxed{9}$ gives a true statement.

$$\begin{array}{r} 8\ 1\ \boxed{\ }\\ +\ \boxed{1}\ \boxed{\ }\ 5\\ \hline \boxed{9}\ 4\ \boxed{\ } \end{array}$$

In the ones column, either $\boxed{3}+5=\boxed{8}$ or $\boxed{8}+5=1\boxed{3}$. Since we must place the 3 and the 8 in the ones column, we can only place the 2 in the tens column.

$$\begin{array}{r} 8\ 1\ \boxed{\ }\\ +\ \boxed{1}\ 2\ 5\\ \hline \boxed{9}\ 4\ \boxed{\ } \end{array}$$

In the tens column, only $1+1+2$ gives a sum of 4. This means the sum of the ones column must be 10 or more. So, the ones column is $\boxed{8}+5=1\boxed{3}$.

$$\begin{array}{r} {\scriptstyle 1}\\ 8\ 1\ \boxed{8}\\ +\ \boxed{1}\ 2\ 5\\ \hline \boxed{9}\ 4\ \boxed{3} \end{array}$$

62. In the ones column, only $\boxed{3}+1$ gives a sum with ones digit 4.

$$\begin{array}{r} \boxed{\ },6\ 2\ \boxed{3}\\ +\ \ 2,7\ \boxed{\ }\ 1\\ \hline 1\ \boxed{\ },4\ \boxed{\ }\ 4 \end{array}$$

In the tens column, either $2+\boxed{0}=\boxed{2}$ or $2+\boxed{8}=1\boxed{0}$.

In the hundreds column, only $1+6+7=14$. So, the hundreds column needs 1 more ten from the tens column.

$$\begin{array}{r} {\scriptstyle 1\ \ 1}\\ \boxed{\ },6\ 2\ \boxed{3}\\ +\ \ 2,7\ \boxed{8}\ 1\\ \hline 1\ \boxed{\ },4\ \boxed{0}\ 4 \end{array}$$

So, the tens column is $2+\boxed{8}=1\boxed{0}$.
The remaining digits are 2 and 9.

In the thousands column, only $1+\boxed{9}+2=1\boxed{2}$ gives a true statement.

$$\begin{array}{r} {\scriptstyle 1\ \ 1}\\ \boxed{9},6\ 2\ \boxed{3}\\ +\ \ 2,7\ \boxed{8}\ 1\\ \hline 1\ \boxed{2},4\ \boxed{0}\ 4 \end{array}$$

63. In the ones column, only $6+\boxed{0}+6=1\boxed{2}$ gives a true statement.

$$\begin{array}{r} {\scriptstyle 1}\\ 5\ \boxed{\ }\ 6\\ 1\ 5\ \boxed{0}\\ +\ \ \boxed{\ }\ 4\ 6\\ \hline \boxed{\ },4\ 6\ \boxed{2} \end{array}$$

In the tens column, only $1+\boxed{6}+5+4=16$ gives a sum that ends in 6.

$$\begin{array}{r} {\scriptstyle 1\ \ 1}\\ 5\ \boxed{6}\ 6\\ 1\ 5\ \boxed{0}\\ +\ \ \boxed{\ }\ 4\ 6\\ \hline \boxed{\ },4\ 6\ \boxed{2} \end{array}$$

The remaining digits are 1 and 7.

In the hundreds column, only $1+5+1+\boxed{7}=\boxed{1}4$ gives a true statement.

$$\begin{array}{r} {\scriptstyle 1\ \ 1}\\ 5\ \boxed{6}\ 6\\ 1\ 5\ \boxed{0}\\ +\ \ \boxed{7}\ 4\ 6\\ \hline \boxed{1},4\ 6\ \boxed{2} \end{array}$$

64. In the ones column, only $9+\boxed{4}+1=14$ gives a sum that ends in 4.

$$\begin{array}{r} {\scriptstyle 1}\\ \boxed{\ },0\ 2\ 9\\ 6,1\ 9\ \boxed{4}\\ +\ \ 1,1\ \boxed{\ }\ 1\\ \hline 1\ \boxed{\ },4\ \boxed{\ }\ 4 \end{array}$$

In the hundreds column, we need 2 more hundreds from the tens column to give $2+0+1+1=4$.

In the tens column, only $1+2+9+\boxed{8}=2\boxed{0}$ gives a sum that is 20 or more.

$$\begin{array}{r} {\scriptstyle 2\ \ 1}\\ \boxed{\ },0\ 2\ 9\\ 6,1\ 9\ \boxed{4}\\ +\ \ 1,1\ \boxed{8}\ 1\\ \hline 1\ \boxed{\ },4\ \boxed{0}\ 4 \end{array}$$

The remaining digits are 2 and 5.

In the thousands column, only $\boxed{5}+6+1=1\boxed{2}$ gives a true statement.

$$\begin{array}{r} {\scriptstyle 2\ \ 1}\\ \boxed{5},0\ 2\ 9\\ 6,1\ 9\ \boxed{4}\\ +\ \ 1,1\ \boxed{8}\ 1\\ \hline 1\ \boxed{2},4\ \boxed{0}\ 4 \end{array}$$

ALGORITHMS
Cross-Sums
52-55

65. In the ones column of the right sum, we have $1+\boxed{4}=5$.

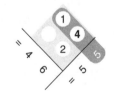

The only remaining digit is 3, so we place it as shown.

Check:
$31+24=55.$ ✓
$14+32=46.$ ✓

66. Step 1:   Step 2:   Final:

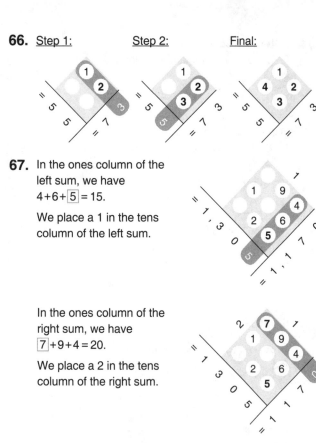

Step 3:   Final:

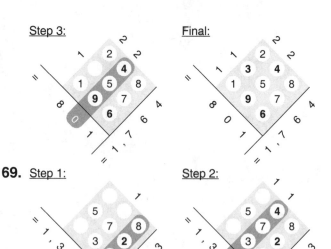

67. In the ones column of the left sum, we have 4+6+$\boxed{5}$ = 15.

We place a 1 in the tens column of the left sum.

In the ones column of the right sum, we have $\boxed{7}$+9+4 = 20.

We place a 2 in the tens column of the right sum.

In the tens column of the right sum, we have 2+1+$\boxed{8}$+6 = 17.

We place a 1 in the hundreds column of the right sum.

The only remaining digit is 3, so we place it as shown.

68. Step 1:   Step 2:

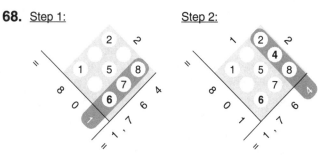

69. Step 1:   Step 2:

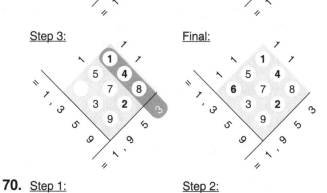

Step 3:   Final:

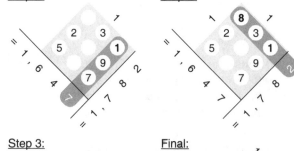

70. Step 1:   Step 2:

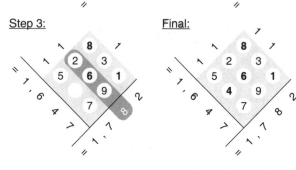

Step 3:   Final:

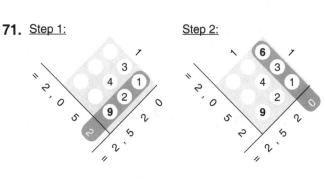

71. Step 1:   Step 2:

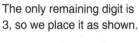

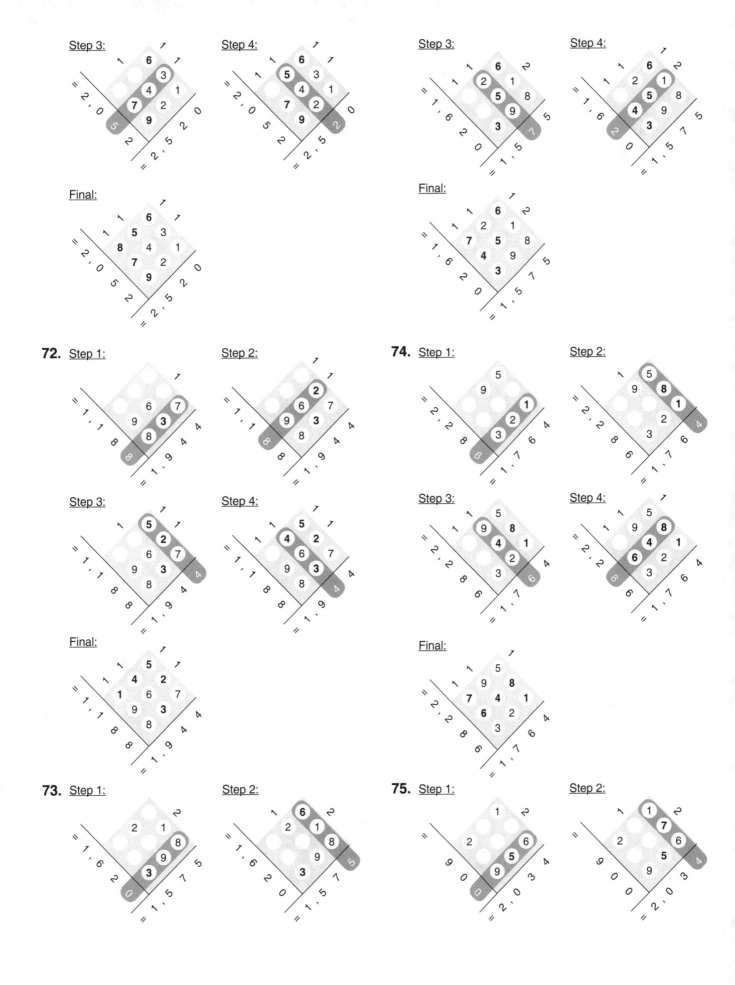

Step 3:

The remaining digits are 3, 4, and 8.

The sum of the tens column of the right sum is between $1+\boxed{3}+\boxed{4}+5 = 13$ and $1+\boxed{8}+\boxed{4}+5 = 18$.

So, we place a 1 in the hundreds column of the right sum.

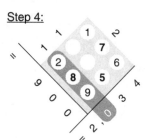

Step 4:

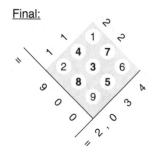

Step 5:

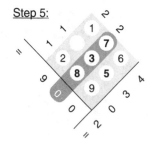

Final:

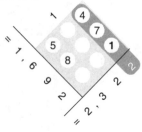

76. Step 1:

Step 2:

The remaining digits are 2, 3, 6, and 9.

In the hundreds column of the right sum, only $1+5+8+\boxed{9} = 23$.

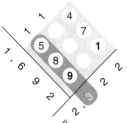

Step 3:

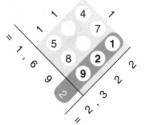

Step 4:

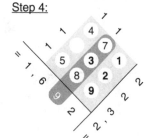

Final:

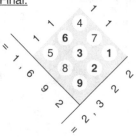

77. Step 1:

The remaining digits are 2, 4, 5, and 8.

In the hundreds column of the left sum, we have $\boxed{}+6+1 = 10$ or $1+\boxed{}+6+1 = 10$.

Only $1+\boxed{2}+6+1 = 10$ gives a true statement.

Step 2:
Step 3:

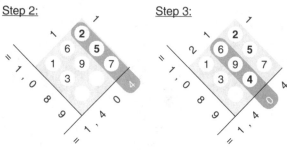

Final:

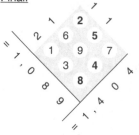

78.

In the ones column of the left sum, only $9+\boxed{2}+\boxed{4} = 15$ has a sum that ends in 5. So, we must place 2 and 4 in this column.

In the ones column of the right sum, only $3+\boxed{4}+\boxed{8} = 15$ has a sum that ends in 5. So, we must place 4 and 8 in this column.

Since the rightmost circle is in both columns, we place the 4 in this circle.

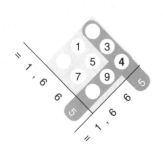

Then, we complete the puzzle as shown.

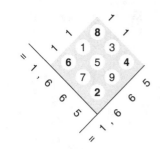

79. In the ones column, we can't take 6 away from 3. So, we break a ten to make 10 more ones.

$$
\begin{array}{r}
{}^{6}\cancel{7}\ {}^{13}\cancel{3} \\
-\ 3\ 6 \\
\hline

\end{array}
$$

7 tens and 3 ones is the same as 6 tens and 13 ones.

Then we can subtract by place value as shown.

$$
\begin{array}{r}
{}^{6}\cancel{7}\ {}^{13}\cancel{3} \\
-\ 3\ 6 \\
\hline
3\ 7
\end{array}
$$

80. In the ones column, we have 7 − 5 = 2.

$$
\begin{array}{r}
5\ 4\ 7 \\
-\ \ 6\ 5 \\
\hline
2
\end{array}
$$

In the tens column, we can't take 6 away from 4. So, we break a hundred to make 10 more tens.

$$
\begin{array}{r}
{}^{4}\cancel{5}\ {}^{14}\cancel{4}\ 7 \\
-\ \ 6\ 5 \\
\hline
2
\end{array}
$$

5 hundreds and 4 tens is the same as 4 hundreds and 14 tens.

Then we can subtract by place value as shown.

$$
\begin{array}{r}
{}^{4}\cancel{5}\ {}^{14}\cancel{4}\ 7 \\
-\ \ 6\ 5 \\
\hline
4\ 8\ 2
\end{array}
$$

81. Step 1:
We break 1 ten into 10 ones.

$$
\begin{array}{r}
{}^{7}\ {}^{12} \\
3\ \cancel{8}\ \cancel{2} \\
-\ \ 5\ 4 \\
\hline

\end{array}
$$

Final:
We subtract by place value.

$$
\begin{array}{r}
{}^{7}\ {}^{12} \\
3\ \cancel{8}\ \cancel{2} \\
-\ \ 5\ 4 \\
\hline
3\ 2\ 8
\end{array}
$$

82. Step 1:
We subtract the ones.

$$
\begin{array}{r}
9\ 2\ 7 \\
-\ 1\ 5\ 5 \\
\hline
2
\end{array}
$$

Step 2:
We break 1 hundred into 10 tens.

$$
\begin{array}{r}
{}^{8}\ {}^{12} \\
\cancel{9}\ \cancel{2}\ 7 \\
-\ 1\ 5\ 5 \\
\hline
2
\end{array}
$$

Final:
We subtract by place value.

$$
\begin{array}{r}
{}^{8}\ {}^{12} \\
\cancel{9}\ \cancel{2}\ 7 \\
-\ 1\ 5\ 5 \\
\hline
7\ 7\ 2
\end{array}
$$

83. Step 1:
We break 1 ten into 10 ones.

$$
\begin{array}{r}
{}^{4}\ {}^{18} \\
7\ \cancel{5}\ \cancel{8} \\
-\ 4\ 3\ 9 \\
\hline

\end{array}
$$

Final:
We subtract by place value.

$$
\begin{array}{r}
{}^{4}\ {}^{18} \\
7\ \cancel{5}\ \cancel{8} \\
-\ 4\ 3\ 9 \\
\hline
3\ 1\ 9
\end{array}
$$

84. Step 1:
We subtract the ones.

$$
\begin{array}{r}
8\ 0\ 7 \\
-\ 3\ 6\ 4 \\
\hline
3
\end{array}
$$

Step 2:
We break 1 hundred into 10 tens.

$$
\begin{array}{r}
{}^{7}\ {}^{10} \\
\cancel{8}\ \cancel{0}\ 7 \\
-\ 3\ 6\ 4 \\
\hline
3
\end{array}
$$

Final:
We subtract by place value.

$$
\begin{array}{r}
{}^{7}\ {}^{10} \\
\cancel{8}\ \cancel{0}\ 7 \\
-\ 3\ 6\ 4 \\
\hline
4\ 4\ 3
\end{array}
$$

85. Step 1:
We subtract the ones and tens.

$$
\begin{array}{r}
5\ ,6\ 4\ 3 \\
-\ 1\ ,9\ 2\ 1 \\
\hline
,2\ 2
\end{array}
$$

Step 2:
We break 1 thousand into 10 hundreds.

$$
\begin{array}{r}
{}^{4}\ {}^{16} \\
\cancel{5}\ ,\cancel{6}\ 4\ 3 \\
-\ 1\ ,9\ 2\ 1 \\
\hline
,2\ 2
\end{array}
$$

Final:
We subtract by place value.

$$
\begin{array}{r}
{}^{4}\ {}^{16} \\
\cancel{5}\ ,\cancel{6}\ 4\ 3 \\
-\ 1\ ,9\ 2\ 1 \\
\hline
3\ ,7\ 2\ 2
\end{array}
$$

86. Step 1:
We subtract the ones.

$$
\begin{array}{r}
9\ ,3\ 8\ 6 \\
-\ 7\ ,1\ 9\ 6 \\
\hline
,0
\end{array}
$$

Step 2:
We break 1 hundred into 10 tens.

$$
\begin{array}{r}
{}^{2}\ {}^{18} \\
9\ ,\cancel{3}\ \cancel{8}\ 6 \\
-\ 7\ ,1\ 9\ 6 \\
\hline
,0
\end{array}
$$

Final:
We subtract by place value.

$$
\begin{array}{r}
{}^{2}\ {}^{18} \\
9\ ,\cancel{3}\ \cancel{8}\ 6 \\
-\ 7\ ,1\ 9\ 6 \\
\hline
2\ ,1\ 9\ 0
\end{array}
$$

87. Step 1:
We break a ten.

$$
\begin{array}{r}
{}^{4}\ {}^{17} \\
8\ \cancel{5}\ \cancel{7} \\
-\ 3\ 6\ 9 \\
\hline

\end{array}
$$

Step 2:
We subtract the ones.

$$
\begin{array}{r}
{}^{4}\ {}^{17} \\
8\ \cancel{5}\ \cancel{7} \\
-\ 3\ 6\ 9 \\
\hline
8
\end{array}
$$

Step 3:
We break a hundred.

$$
\begin{array}{r}
{}^{14} \\
{}^{7}\ \cancel{4}\ {}^{17} \\
\cancel{8}\ \cancel{5}\ \cancel{7} \\
-\ 3\ 6\ 9 \\
\hline
8
\end{array}
$$

Final:
Subtract by place value.

$$
\begin{array}{r}
{}^{14} \\
{}^{7}\ \cancel{4}\ {}^{17} \\
\cancel{8}\ \cancel{5}\ \cancel{7} \\
-\ 3\ 6\ 9 \\
\hline
4\ 8\ 8
\end{array}
$$

88. Step 1:
We break a ten.

$$
\begin{array}{r}
{}^{0}\ {}^{14} \\
5\ \cancel{1}\ \cancel{4} \\
-\ 3\ 8\ 6 \\
\hline

\end{array}
$$

Step 2:
We subtract the ones.

$$
\begin{array}{r}
{}^{0}\ {}^{14} \\
5\ \cancel{1}\ \cancel{4} \\
-\ 3\ 8\ 6 \\
\hline
8
\end{array}
$$

Step 3:
We break a hundred.

$$
\begin{array}{r}
{}^{10} \\
{}^{4}\ \cancel{0}\ {}^{14} \\
\cancel{5}\ \cancel{1}\ \cancel{4} \\
-\ 3\ 8\ 6 \\
\hline
8
\end{array}
$$

Final:
Subtract by place value.

$$
\begin{array}{r}
{}^{10} \\
{}^{4}\ \cancel{0}\ {}^{14} \\
\cancel{5}\ \cancel{1}\ \cancel{4} \\
-\ 3\ 8\ 6 \\
\hline
1\ 2\ 8
\end{array}
$$

89. In the ones place, we can't take 9 away from 3. So, we need to break a ten to make 10 more ones.

$$\begin{array}{r} {\scriptstyle 1\ 10} \\ 2\ \cancel{0}\ 3 \\ -\quad\ 5\ 9 \\ \hline \square\square\square \end{array}$$

But, there are 0 tens in 203. So, we first break a hundred to make 10 more tens. 2 hundreds and 0 tens is the same as 1 hundred and 10 tens.

Now we can break 1 ten to make 10 ones. 10 tens and 3 ones is the same as 9 tens and 13 ones.

$$\begin{array}{r} {\scriptstyle 9} \\ {\scriptstyle 1\ \cancel{10}\ 13} \\ \cancel{2}\ \cancel{0}\ \cancel{3} \\ -\quad\ 5\ 9 \\ \hline \square\square\square \end{array}$$

Finally, we subtract by place value.

In the ones column, $13-9=4$.
In the tens column, $9-5=4$.
In the hundreds column, we have 1 hundred.

$$\begin{array}{r} {\scriptstyle 9} \\ {\scriptstyle 1\ \cancel{10}\ 13} \\ \cancel{2}\ \cancel{0}\ \cancel{3} \\ -\quad\ 5\ 9 \\ \hline \boxed{1}\ \boxed{4}\ \boxed{4} \end{array}$$

90. First we stack the numbers. Then, we subtract.

Step 1: We break a ten.

$$\begin{array}{r} {\scriptstyle 3\ 13} \\ \boxed{4}\ \boxed{4}\ \boxed{3} \\ -\quad \boxed{6}\ \boxed{7} \\ \hline \square\square\square \end{array}$$

Step 2: We subtract the ones.

$$\begin{array}{r} {\scriptstyle 3\ 13} \\ 4\ 4\ 3 \\ -\quad 6\ 7 \\ \hline \square\square\ 6 \end{array}$$

Step 3: We break a hundred.

$$\begin{array}{r} {\scriptstyle 13} \\ {\scriptstyle 3\ \cancel{3}\ 13} \\ 4\ 4\ 3 \\ -\quad 6\ 7 \\ \hline \square\ 6 \end{array}$$

Final: Subtract by place value.

$$\begin{array}{r} {\scriptstyle 13} \\ {\scriptstyle 3\ \cancel{3}\ 13} \\ 4\ 4\ 3 \\ -\quad 6\ 7 \\ \hline 3\ 7\ 6 \end{array}$$

So, $443-67 = \mathbf{376}$.

91. First we stack the numbers. Then, we subtract.

To break a ten, we must first break a hundred.

Step 1: We break a hundred.

$$\begin{array}{r} {\scriptstyle 7\ 10} \\ \boxed{5},\boxed{8}\boxed{0}\boxed{6} \\ -\quad \boxed{5}\boxed{6}\boxed{7} \\ \hline \square,\square\square\square \end{array}$$

Step 2: We break a ten.

$$\begin{array}{r} {\scriptstyle 9} \\ {\scriptstyle 7\ \cancel{10}\ 16} \\ 5,8\ 0\ 6 \\ -\quad 5\ 6\ 7 \\ \hline \square,\square\square\square \end{array}$$

Final: Subtract by place value.

$$\begin{array}{r} {\scriptstyle 9} \\ {\scriptstyle 7\ \cancel{10}\ 16} \\ 5,8\ 0\ 6 \\ -\quad 5\ 6\ 7 \\ \hline 5,2\ 3\ 9 \end{array}$$

So, $5,806-567 = \mathbf{5,239}$.

92. First we stack the numbers. Then, we subtract.

Step 1: We break a ten.

$$\begin{array}{r} {\scriptstyle 0\ 16} \\ 9\ \cancel{1}\ 6 \\ -\ 2\ 4\ 8 \\ \hline \end{array}$$

Step 2: We subtract the ones.

$$\begin{array}{r} {\scriptstyle 0\ 16} \\ 9\ \cancel{1}\ 6 \\ -\ 2\ 4\ 8 \\ \hline 8 \end{array}$$

Step 3: We break a hundred.

$$\begin{array}{r} {\scriptstyle 10} \\ {\scriptstyle 8\ \cancel{0}\ 16} \\ \cancel{9}\ \cancel{1}\ 6 \\ -\ 2\ 4\ 8 \\ \hline 8 \end{array}$$

Final: Subtract by place value.

$$\begin{array}{r} {\scriptstyle 10} \\ {\scriptstyle 8\ \cancel{0}\ 16} \\ \cancel{9}\ \cancel{1}\ 6 \\ -\ 2\ 4\ 8 \\ \hline 6\ 6\ 8 \end{array}$$

So, $916-248 = \mathbf{668}$.

93. First we stack the numbers. Then, we subtract.

Step 1: We break a ten.

$$\begin{array}{r} {\scriptstyle 2\ 16} \\ 2,7\ \cancel{3}\ \cancel{6} \\ -\quad 8\ 2\ 9 \\ \hline \end{array}$$

Step 2: We subtract the ones and tens.

$$\begin{array}{r} {\scriptstyle 2\ 16} \\ 2,7\ \cancel{3}\ \cancel{6} \\ -\quad 8\ 2\ 9 \\ \hline 0\ 7 \end{array}$$

Step 3: We break a thousand.

$$\begin{array}{r} {\scriptstyle 1\ 17\ 2\ 16} \\ \cancel{2},\cancel{7}\ \cancel{3}\ \cancel{6} \\ -\quad 8\ 2\ 9 \\ \hline 0\ 7 \end{array}$$

Final: We subtract by place value.

$$\begin{array}{r} {\scriptstyle 1\ 17\ 2\ 16} \\ \cancel{2},\cancel{7}\ \cancel{3}\ \cancel{6} \\ -\quad 8\ 2\ 9 \\ \hline 1,9\ 0\ 7 \end{array}$$

So, $2,736-829 = \mathbf{1,907}$.

94. First we stack the numbers. Then, we subtract.

Step 1: We break a ten.

$$\begin{array}{r} {\scriptstyle 5\ 17} \\ 3,0\ 6\ \cancel{7} \\ -\ 2,2\ 7\ 8 \\ \hline \end{array}$$

Step 2: We subtract the ones.

$$\begin{array}{r} {\scriptstyle 5\ 17} \\ 3,0\ 6\ \cancel{7} \\ -\ 2,2\ 7\ 8 \\ \hline 9 \end{array}$$

To break a hundred, we must first break a thousand.

Step 3: We break a thousand.

$$\begin{array}{r} {\scriptstyle 2\ 10\ 5\ 17} \\ \cancel{3},\cancel{0}\ 6\ \cancel{7} \\ -\ 2,2\ 7\ 8 \\ \hline 9 \end{array}$$

Step 4: We break a hundred.

$$\begin{array}{r} {\scriptstyle 9\ 15} \\ {\scriptstyle 2\ \cancel{10}\ \cancel{5}\ 17} \\ \cancel{3},\cancel{0}\ 6\ \cancel{7} \\ -\ 2,2\ 7\ 8 \\ \hline 9 \end{array}$$

Final: We subtract by place value.

$$\begin{array}{r} {\scriptstyle 9\ 15} \\ {\scriptstyle 2\ \cancel{10}\ \cancel{5}\ 17} \\ \cancel{3},\cancel{0}\ 6\ \cancel{7} \\ -\ 2,2\ 7\ 8 \\ \hline 7\ 8\ 9 \end{array}$$

So, $3,067-2,278 = \mathbf{789}$.

95. First we stack the numbers. Then, we subtract.

Step 1: We break a ten.

$$\begin{array}{r} {\scriptstyle 5\ 11} \\ 4\ 0,0\ 6\ \cancel{1} \\ -\ 1\ 3,4\ 5\ 6 \\ \hline \end{array}$$

Step 2: We subtract the ones and tens.

$$\begin{array}{r} {\scriptstyle 5\ 11} \\ 4\ 0,0\ 6\ \cancel{1} \\ -\ 1\ 3,4\ 5\ 6 \\ \hline 0\ 5 \end{array}$$

To break a thousand, we must first break a ten-thousand.

Step 3:
We break
a ten-thousand.

$$
\begin{array}{r}
3\ 10 \quad 5\ 11 \\
\cancel{4}\ \cancel{0}\ ,\ 0\ \cancel{6}\ \cancel{1} \\
-\ 1\ 3\ ,\ 4\ 5\ 6 \\
\hline
0\ 5
\end{array}
$$

Step 4:
We break
a thousand.

$$
\begin{array}{r}
9 \\
3\ \cancel{10}\ 10\ 5\ 11 \\
\cancel{4}\ \cancel{0}\ ,\ 0\ \cancel{6}\ \cancel{1} \\
-\ 1\ 3\ ,\ 4\ 5\ 6 \\
\hline
0\ 5
\end{array}
$$

Final:
We subtract by
place value.

$$
\begin{array}{r}
9 \\
3\ \cancel{10}\ 10\ 5\ 11 \\
\cancel{4}\ \cancel{0}\ ,\ \cancel{0}\ \cancel{6}\ \cancel{1} \\
-\ 1\ 3\ ,\ 4\ 5\ 6 \\
\hline
2\ 6\ ,\ 6\ 0\ 5
\end{array}
$$

So, $40{,}061 - 13{,}456 = \mathbf{26{,}605}$.

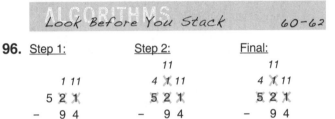
ALGORITHMS
Look Before You Stack   60-62

96. Step 1:

$$
\begin{array}{r}
1\ 11 \\
5\ 2\ \cancel{1} \\
-\ \ 9\ 4 \\
\hline
\end{array}
$$

Step 2:

$$
\begin{array}{r}
11 \\
4\ \cancel{\cancel{1}}\ 11 \\
5\ \cancel{2}\ \cancel{1} \\
-\ \ 9\ 4 \\
\hline
7
\end{array}
$$

Final:

$$
\begin{array}{r}
11 \\
4\ \cancel{\cancel{1}}\ 11 \\
5\ \cancel{2}\ \cancel{1} \\
-\ \ 9\ 4 \\
\hline
4\ 2\ 7
\end{array}
$$

97. Taking away 94 is the same as taking away 100, then adding 6. So,

$$
521 - 94 = 521 - 100 + \mathbf{6}
$$
$$
= \mathbf{427}.
$$

98. Step 1:

$$
\begin{array}{r}
1 \\
3\ 9\ 6 \\
+\ \ 8\ 7 \\
\hline
3
\end{array}
$$

Step 2:

$$
\begin{array}{r}
1\ 1 \\
3\ 9\ 6 \\
+\ \ 8\ 7 \\
\hline
8\ 3
\end{array}
$$

Final:

$$
\begin{array}{r}
1\ 1 \\
3\ 9\ 6 \\
+\ \ 8\ 7 \\
\hline
4\ 8\ 3
\end{array}
$$

99. We take 4 from 87 and give it to 396. So,

$$
396 + 87 = 400 + \mathbf{83}
$$
$$
= \mathbf{483}.
$$

100. Step 1:

$$
\begin{array}{r}
4\ 17 \\
8\ \cancel{5}\ \cancel{7} \\
-\ \ 5\ 9 \\
\hline
\end{array}
$$

Step 2:

$$
\begin{array}{r}
14 \\
7\ \cancel{4}\ 17 \\
8\ \cancel{5}\ \cancel{7} \\
-\ \ 5\ 9 \\
\hline
8
\end{array}
$$

Final:

$$
\begin{array}{r}
14 \\
7\ \cancel{4}\ 17 \\
8\ \cancel{5}\ \cancel{7} \\
-\ \ 5\ 9 \\
\hline
7\ 9\ 8
\end{array}
$$

101. Taking away 59 is the same as taking away 57, then taking away 2.

Taking away 57 from 857 gives 800. Then, taking away 2 more gives 798. So,

$$
857 - 59 = 857 - 57 - \mathbf{2}
$$
$$
= \mathbf{798}.
$$

102. Step 1:

$$
\begin{array}{r}
2 \\
1\ 9\ 7 \\
1\ 9\ 8 \\
+\ 1\ 9\ 9 \\
\hline
4
\end{array}
$$

Step 2:

$$
\begin{array}{r}
2\ 2 \\
1\ 9\ 7 \\
1\ 9\ 8 \\
+\ 1\ 9\ 9 \\
\hline
9\ 4
\end{array}
$$

Final:

$$
\begin{array}{r}
2\ 2 \\
1\ 9\ 7 \\
1\ 9\ 8 \\
+\ 1\ 9\ 9 \\
\hline
5\ 9\ 4
\end{array}
$$

103. 197 is 3 less than 200.
198 is 2 less than 200.
199 is 1 less than 200.

So, $197+198+199$ is $3+2+1 = 6$ less than $200+200+200 = 600$. This means

$$
197 + 198 + 199 = 600 - \mathbf{6}
$$
$$
= \mathbf{594}.
$$

104. Step 1:

$$
\begin{array}{r}
1 \\
2\ ,\ 6\ 7\ 5 \\
+\ \ \ 9\ 9\ 8 \\
\hline
3
\end{array}
$$

Step 2:

$$
\begin{array}{r}
1\ 1 \\
2\ ,\ 6\ 7\ 5 \\
+\ \ \ 9\ 9\ 8 \\
\hline
7\ 3
\end{array}
$$

Step 3:

$$
\begin{array}{r}
1\ 1\ 1 \\
2\ ,\ 6\ 7\ 5 \\
+\ \ \ 9\ 9\ 8 \\
\hline
6\ 7\ 3
\end{array}
$$

Final:

$$
\begin{array}{r}
1\ 1\ 1 \\
2\ ,\ 6\ 7\ 5 \\
+\ \ \ 9\ 9\ 8 \\
\hline
3\ ,\ 6\ 7\ 3
\end{array}
$$

105. We add 1,000, then take away 2. So,

$$
2{,}675 + 998 = 2{,}675 + 1{,}000 - \mathbf{2}
$$
$$
= \mathbf{3{,}673}.
$$

106. To break a ten, we must first break a thousand, then break a hundred.

Step 1:

$$
\begin{array}{r}
0\ 10 \\
\cancel{1}\ ,\ 0\ 0\ 0 \\
-\ \ \ 8\ 8\ 7 \\
\hline
\end{array}
$$

Step 2:

$$
\begin{array}{r}
9 \\
0\ \cancel{10}\ 10 \\
\cancel{1}\ ,\ \cancel{0}\ 0\ 0 \\
-\ \ \ 8\ 8\ 7 \\
\hline
\end{array}
$$

Step 3:

$$
\begin{array}{r}
9\ 9 \\
0\ \cancel{10}\ \cancel{10}\ 10 \\
\cancel{1}\ ,\ \cancel{0}\ \cancel{0}\ 0 \\
-\ \ \ 8\ 8\ 7 \\
\hline
\end{array}
$$

Final:

$$
\begin{array}{r}
9\ 9 \\
0\ \cancel{10}\ \cancel{10}\ 10 \\
\cancel{1}\ ,\ \cancel{0}\ \cancel{0}\ 0 \\
-\ \ \ 8\ 8\ 7 \\
\hline
1\ 1\ 3
\end{array}
$$

107. We count up by 13 to get from 887 to 900.
We count up by 100 to get from 900 to 1,000.

All together, we count up by $13+100 = 113$ to get from 887 to 1,000.

$$
887 + \mathbf{113} = 1{,}000.
$$
$$
\text{So, } 1{,}000 - 887 = \mathbf{113}.
$$

108. To break a ten, we must first break a thousand, then break a hundred.

Step 1:

```
    7 10
  8 , 0 0 2
 - 7 , 9 9 6
```

Step 2:

```
        9
    7 10 10
  8 , 0 0 2
 - 7 , 9 9 6
```

Step 3:

```
      9  9
   7 10 10 12
  8 , 0 0 2
 - 7 , 9 9 6
```

Final:

```
      9  9
   7 10 10 12
  8 , 0 0 2
 - 7 , 9 9 6
          6
```

109. We count up by 6 to get from 7,996 to 8,002.

$$7{,}996 + \underline{6} = 8{,}002.$$
$$\text{So, } 8{,}002 - 7{,}996 = \underline{6}.$$

Notice that sometimes it is a lot easier to add or subtract without stacking!

110. We count up by 4 to get from 99,996 to 100,000.
We count up by 13 to get from 100,000 to 100,013.

All together, we count up by $4+13 = 17$ to get from 99,996 to 100,013. So, $100{,}013 - 99{,}996 = \mathbf{17}$.

111. We take 20 away from 2,220 and give it to 1,980.

So, $1{,}980 + 2{,}220 = 2{,}000 + 2{,}200 = \mathbf{4{,}200}$.

112. Taking away 819 is the same as taking away 817, then taking away 2.

So, $8{,}817 - 819 = 8{,}817 - 817 - 2 = \mathbf{7{,}998}$.

113. We count up by 75 to get from 1,725 to 1,800.
We count up by 200 to get from 1,800 to 2,000.
We count up by 2,000 to get from 2,000 to 4,000.

All together, we count up by $75 + 200 + 2{,}000 = 2{,}275$ to get from 1,725 to 4,000. So, $4{,}000 - 1{,}725 = \mathbf{2{,}275}$.

114. 9,990 is 10 less than 10,000.
9,995 is 5 less than 10,000.

So, $9{,}990 + 9{,}995$ is $10 + 5 = 15$ less than $10{,}000 + 10{,}000 = 20{,}000$.

So, $9{,}990 + 9{,}995 = 20{,}000 - 15 = \mathbf{19{,}985}$.

115. Taking away 596 is the same as taking away 600, then adding 4.

So, $2{,}772 - 596 = 2{,}772 - 600 + 4 = \mathbf{2{,}176}$.

116. To add 2,980, we add 3,000, then take away 20.

So, $3{,}333 + 2{,}980 = 3{,}333 + 3{,}000 - 20 = \mathbf{6{,}313}$.

117. We take 25 from 625 and give it to 3,975.

So, $3{,}975 + 625 = 4{,}000 + 600 = \mathbf{4{,}600}$.

118. We count up by 2 to get from 53,209 to 53,211.

So, $53{,}211 - 53{,}209 = \mathbf{2}$.

119. 9,999 is 1 less than 10,000.
9,998 is 2 less than 10,000.
9,997 is 3 less than 10,000.

So, $9{,}999 + 9{,}998 + 9{,}997$ is $1 + 2 + 3 = 6$ less than $10{,}000 + 10{,}000 + 10{,}000 = 30{,}000$.

So, $9{,}999 + 9{,}998 + 9{,}997 = 30{,}000 - 6 = \mathbf{29{,}994}$.

ALGORITHMS
Checking   63

120. If $529 - 146 = 383$, then $383 + 146$ is 529.
We check the addition as shown to the right.

```
    1
  3 8 3
+ 1 4 6
  5 2 9
```

Since $383 + 146 = 529$, the original subtraction is correct. ✓

If $716 - 344 = 362$, then $362 + 344$ is 716.
We check the addition as shown to the right.

```
    1
  3 6 2
+ 3 4 4
  7 0 6
```

Since $362 + 344 = 7\underline{0}6$, the original subtraction is incorrect. ✗

If $518 - 343 = 175$, then $175 + 343$ is 518.
We check the addition as shown to the right.

```
    1
  1 7 5
+ 3 4 3
  5 1 8
```

Since $175 + 343 = 518$, the original subtraction is correct. ✓

So, we cross out the second subtraction.

```
  5 2 9        7̶ 1̶ 6̶        5 1 8
- 1 4 6      - 3̶ 4̶ 4̶      - 3 4 3
  3 8 3        3̶ 6̶ 2̶        1 7 5
```

121. If $706 - 238 = 458$, then $458 + 238$ is 706.
We check the addition as shown to the right.

```
    1
  4 5 8
+ 2 3 8
  6 9 6
```

Since $458 + 238 = \underline{6}96$, the original subtraction is incorrect. ✗

If $804 - 327 = 477$, then $477 + 327$ is 804.
We check the addition as shown to the right.

```
  1 1
  4 7 7
+ 3 2 7
  8 0 4
```

Since $477 + 327 = 804$, the original subtraction is correct. ✓

If $605 - 346 = 369$, then $369 + 346$ is 605.
We check the addition as shown to the right.

```
  1 1
  3 6 9
+ 3 4 6
  7 1 5
```

Since $369 + 346 = \underline{7}15$, the original subtraction is incorrect. ✗

So, we cross out the first and third subtractions.

```
  7̶ 0̶ 6̶        8 0 4        6̶ 0̶ 5̶
- 2̶ 3̶ 8̶      - 3 2 7      - 3̶ 4̶ 6̶
  4̶ 5̶ 8̶        4 7 7        3̶ 6̶ 9̶
```

122. If $513 - 357 = 156$, then $156 + 357$ is 513.
We check the addition as shown to the right.

```
  1 1
  1 5 6
+ 3 5 7
  5 1 3
```

Since $156 + 357 = 513$, the original subtraction is correct. ✓

If $753 - 175 = 572$, then $572 + 175$ is 753.
We check the addition as shown to the right.

```
    1
  5 7 2
+ 1 7 5
  7 4 7
```

Since $572 + 175 = 7\underline{4}7$, the original subtraction is incorrect. ✗

If $931-577=354$, then $354+577$ is 931. We check the addition as shown to the right.

Since $354+577=931$, the original subtraction is correct. ✓

```
    1 1
    3 5 4
  + 5 7 7
    9 3 1
```

So, we cross out the second subtraction.

```
  5 1 3        7̶ 5̶ 3̶       9 3 1
- 3 5 7      - 1̶ 7̶ 5̶     - 5 7 7
  1 5 6        5̶ 7̶ 2̶       3 5 4
```

123. Mount Scylla is $4,184-3,427=$ **757** meters taller than Mount Cyclops.

```
    3 11 7 14
    4̶,1̶ 8̶ 4̶
  - 3,4 2 7
      7 5 7
```

124. $1,344-576=$ **768** of the eggs survived.

```
          12 13
      0 2̶ 3̶ 14
      1̶,3̶ 4̶ 4̶
  -     5 7 6
        7 6 8
```

125. 1,478 years before 2,589 years ago, the Behemoth Pine sprouted. So, the Behemoth Pine is $1,478+2,589=$ **4,067** years old now.

```
    1   1 1
    1,4 7 8
  + 2,5 8 9
    4,0 6 7
```

126. There are $6,435-1,365=$ **5,070** more possible 7-topping pizzas than 4-topping pizzas.

```
      3 13
    6,4̶ 3̶ 5
  - 1,3 6 5
    5,0 7 0
```

127. Three boxes of Beastie Bites cereal have $567+567+567=$ **1,701** grams of cereal.

```
    2 2
    5 6 7
    5 6 7
  + 5 6 7
    1,7 0 1
```

128. The trip from Hydra Harbor to Cape Capricorn and back is $2,250+2,250=4,500$ miles by boat.

```
        1
    2,2 5 0
  + 2,2 5 0
    4,5 0 0
```

The trip from Hydra Harbor to Cape Capricorn and back is $2,827+2,827=5,654$ miles by car.

```
    1   1
    2,8 2 7
  + 2,8 2 7
    5,6 5 4
```

So, the trip from Hydra Harbor to Cape Capricorn and back is $5,654-4,500=$ **1,154** miles shorter by boat than by car.

```
    5,6 5 4
  - 4,5 0 0
    1,1 5 4
```

— *or* —

The trip from Hydra Harbor to Cape Capricorn is $2,827-2,250=577$ miles shorter by boat than by car.

```
      7 12
    2,8̶ 2̶ 7
  - 2,2 5 0
      5 7 7
```

So, the trip from Hydra Harbor to Cape Capricorn *and back* is $577+577=$ **1,154** miles shorter by boat than by car.

```
    1 1
    5 7 7
  + 5 7 7
    1,1 5 4
```

129. Cutlass Pete's ship weighs $1,456+300=1,756$ tons. So, the two ships together weigh $1,456+1,756=$ **3,212** tons.

```
    1 1 1
    1,4 5 6
  + 1,7 5 6
    3,2 1 2
```

130. Since 2,156 of Borg's 3,000 cards are in a shoebox, the other $3,000-2,156=844$ cards are in the carrying case.

```
        9 9
    2 1̶0 1̶0 10
    3̶,0̶ 0̶ 0̶
  - 2,1 5 6
        8 4 4
```

So, there are $2,156-844=$ **1,312** more cards in Borg's shoebox than in his carrying case.

```
    1 11
    2,1̶ 5 6
  -   8 4 4
    1,3 1 2
```

131. In the ones column, $2+2=\boxed{4}$. So, A is 4. We replace each A with 4.

```
      4 2
    + 4 2
    ⊡ 4
```

In the tens column, $4+4=\boxed{8}$. So, B is 8. We have A = **4** and B = **8**.

```
      4 2
    + 4 2
    8 4
```

132. In the ones column, $5+\boxed{3}=8$. So, A is 3. We replace each A with 3.

```
      3 5
    + 4 3
    ⊡ 8
```

In the tens column, $3+4=\boxed{7}$. So, B is 7. We have A = **3** and B = **7**.

```
      3 5
    + 4 3
    7 8
```

133. In the ones column, $5+5=1\boxed{0}$. So, A is 0.

We replace each A with 0 and place a 1 above the tens column.

```
        1
    ⊡ 0 5
  + ⊡ 6 5
    9 ⊡ 0
```

In the tens column, $1+0+6=\boxed{7}$. So, B is 7.

We replace each B with 7.

```
        1
    7 0 5
  + ⊡ 6 5
    9 7 0
```

In the hundreds column, $7+\boxed{2}=9$.
So, C is 2.

We have A = **0**, B = **7**, and C = **2**.

```
      1
    7 0 5
  + 2 6 5
  ───────
    9 7 0
```

134. In the ones column, $\boxed{1}+2=3$. So, A is 1.
We replace each A with 1.

```
  Ⓑ 1 1
+ Ⓑ 4 2
─────────
1 , Ⓒ Ⓑ 3
```

In the tens column, $1+4=\boxed{5}$. So, B is 5.
We replace each B with 5.

```
  5 1 1
+ 5 4 2
─────────
1 , Ⓒ 5 3
```

In the hundreds column, $5+5=1\boxed{0}$.
So, C is 0.

We have A = **1**, B = **5**, and C = **0**.

```
  5 1 1
+ 5 4 2
─────────
1 , 0 5 3
```

135. In the ones column, $8+8=1\boxed{6}$.
So, A is 6.

We replace each A with 6 and place a 1
above the tens column.

```
      1
  Ⓑ 6 8
+ Ⓑ 6 8
───────
  Ⓒ Ⓑ 6
```

In the tens column, $1+6+6=1\boxed{3}$.
So, B is 3.

We replace each B with 3 and place a 1
above the hundreds column.

```
    1 1
  3 6 8
+ 3 6 8
───────
  Ⓒ 3 6
```

In the hundreds column, $1+3+3=\boxed{7}$.
So, C is 7.

We have A = **6**, B = **3**, and C = **7**.

```
    1 1
  3 6 8
+ 3 6 8
───────
  7 3 6
```

136. In the ones column, $3+5=\boxed{8}$.
So, A is 8.

We replace each A with 8.

```
  8 Ⓑ 3
+ 8 5 5
─────────
Ⓑ , Ⓒ Ⓒ 8
```

The largest possible sum of two 3-digit
numbers is $999+999=1,998$. So, the
thousands digit of the sum must be 1.

So, B is 1. We replace each B with 1.

```
  8 1 3
+ 8 5 5
─────────
1 , Ⓒ Ⓒ 8
```

In the tens column, $1+5=\boxed{6}$.
So, C is 6.

We have A = **8**, B = **1**, and C = **6**.

```
  8 1 3
+ 8 5 5
─────────
1 , 6 6 8
```

137. In the ones column, only adding 0 to a
digit gives the same digit. So, B is 0.

We replace each B with 0.

```
  Ⓐ Ⓐ
+ Ⓐ 0
───────
Ⓒ 0 Ⓐ
```

In the tens column, only $\boxed{5}+\boxed{5}=1\boxed{0}$.
So, A is 5 and C is 1.

We have A = **5**, B = **0**, and C = **1**.

```
  5 5
+ 5 0
───────
1 0 5
```

138. The largest possible sum of two 3-digit
numbers is $999+999=1,998$. So, the
thousands digit of the sum must be 1.

So, D is 1. We replace each D with 1.

```
  Ⓑ Ⓐ Ⓑ
+ Ⓑ Ⓑ Ⓑ
─────────
1 , 1 Ⓒ Ⓒ
```

In the hundreds column, we cannot add
a number to itself and get 11. But,
$1+\boxed{5}+\boxed{5}=11$. So, there is a 1 above the
hundreds column and B is 5. We replace
each B with 5.

```
      1
  5 Ⓐ 5
+ 5 5 5
─────────
1 , 1 Ⓒ Ⓒ
```

In the ones column, $5+5=1\boxed{0}$. So, C
is 0, and we place a 1 above the tens
column. We replace each C with 0.

```
    1 1
  5 Ⓐ 5
+ 5 5 5
─────────
1 , 1 0 0
```

In the tens column, we have
$1+\boxed{4}+5=10$. So, A is 4.

We have A = **4**, B = **5**, C = **0**, and D = **1**.

```
    1 1
  5 4 5
+ 5 5 5
─────────
1 , 1 0 0
```

139. The smallest possible five-digit result is 10,000. So,
we are looking for the number we add to 8,325 to get
10,000.

To find the number we add to 8,325 to get 10,000, we
subtract $10,000-8,325$.

```
          9 9 9
    0 10 10 10 10
    1̶ 0̶ , 0̶ 0̶ 0̶
  −   8 , 3 2 5
  ─────────────
      1 , 6 7 5
```

So, **1,675** is the smallest number that can be added to
8,325 to give a five-digit result.

— or —

We count up from 8,325 to 10,000.

Adding 75 to 8,325 gives 8,400.
Adding 600 to 8,400 gives 9,000.
Adding 1,000 to 9,000 gives 10,000.

So, $75+600+1,000=$ **1,675** is the smallest number that
can be added to 8,325 to give a five-digit result.

140. To find the missing number in a subtraction problem, we
can rewrite it.

For example, to find the missing number in $7-\_\_=4$, we
can use the equation $7-4=\_\_$.

So, to find the missing number in
$123,456-_____=45,678$, we subtract
$123,456-45,678=_____$.

```
      11 12 13 14
    0  1̶  2̶  3̶  4̶  16
    1̶  2 3 , 4 5 6̶
  −       4 5 , 6 7 8
  ───────────────────
          7 7 , 7 7 8
```

Since $123,456-45,678=77,778$, we have
$123,456-$ **77,778** $=45,678$.

141. 1 number has a 1 in the hundred-thousands place.
2 numbers have a 1 in the ten-thousands place.
3 numbers have a 1 in the thousands place.
4 numbers have a 1 in the hundreds place.

5 numbers have a 1 in the tens place.
6 numbers have a 1 in the ones place.

All together, we have 1 hundred-thousand,
2 ten-thousands, 3 thousands, 4 hundreds, 5 tens,
and 6 ones.

So, $1+11+111+1,111+11,111+111,111 = \mathbf{123,456}$.

142. We stack all five numbers and add.

$$
\begin{array}{r}
2\ 3\ 3\ \ \ \\
2,3\ 4\ 5\\
3,4\ 5\ 6\\
4,5\ 6\ 7\\
5,6\ 7\ 8\\
+\ 6,7\ 8\ 9\\
\hline
2\ 2,8\ 3\ 5
\end{array}
$$

So, $2,345+3,456+4,567+5,678+6,789 = \mathbf{22,835}$.

— *or* —

We know that
$1,234+2,345+3,456+4,567+5,678 = 17,280$.

$2,345+3,456+4,567+5,678+6,789$ is the same sum,
but with 1,234 replaced by 6,789. So the new sum is
$6,789-1,234 = 5,555$ greater than 17,280.

We compute $17,280+5,555$ as shown below.

$$
\begin{array}{r}
1\ \ \ \ 1\ \ \ \ \\
1\ 7,2\ 8\ 0\\
+\ 5,5\ 5\ 5\\
\hline
2\ 2,8\ 3\ 5
\end{array}
$$

So, $2,345+3,456+4,567+5,678+6,789 = \mathbf{22,835}$.

— *or* —

We compare the two sums,
$1,234+2,345+3,456+4,567+5,678$ and
$2,345+3,456+4,567+5,678+6,789$.

Each number in the second sum is 1,111 more than a
number in the first sum. So, the second sum is
$1,111+1,111+1,111+1,111+1,111 = 5,555$ more than
the first sum.

Since the first sum is 17,280, the second sum is
$17,280+5,555$. We compute $17,280+5,555$ as shown
below.

$$
\begin{array}{r}
1\ \ \ \ 1\ \ \ \ \\
1\ 7,2\ 8\ 0\\
+\ 5,5\ 5\ 5\\
\hline
2\ 2,8\ 3\ 5
\end{array}
$$

So, $2,345+3,456+4,567+5,678+6,789 = \mathbf{22,835}$.

143. In the ones column, we have that 8 minus a digit gives
that same digit, or 18 minus a digit gives that same digit.
The only digits that work are $8-\boxed{4}=\boxed{4}$ or $18-\boxed{9}=\boxed{9}$.

If each blank is 4, then we have $14,448-4,444 = 4,444$,
which is not true. So, each blank cannot be 4.

If each blank is 9, then we have

$19,998-9,999 = 19,998-10,000+1 = 9,999$. This works,
so the missing digit is 9. We fill each box with 9.

$$
\begin{array}{r}
1\ \boxed{9},\boxed{9}\ \boxed{9}\ 8\\
-\ \ \ \boxed{9},\boxed{9}\ \boxed{9}\ 9\\
\hline
\boxed{9},\boxed{9}\ \boxed{9}\ \boxed{9}
\end{array}
$$

144. We try adding some numbers that only use the digit 5.

$$
\begin{array}{r}
1\ \ \ \ \\
5\ 5\\
+\ \ \ 5\\
\hline
6\ 0
\end{array}
\qquad
\begin{array}{r}
1\ \ \ \ \ \\
5\ 5\ 5\\
+\ \ \ \ \ 5\\
\hline
5\ 6\ 0
\end{array}
\qquad
\begin{array}{r}
1\ 1\ \ \ \\
5\ 5\ 5\\
+\ \ \ 5\ 5\\
\hline
6\ 1\ 0
\end{array}
\qquad
\begin{array}{r}
1\ 1\ \ \ \ \\
5,5\ 5\ 5\\
+\ \ \ \ \ 5\ 5\\
\hline
5,6\ 1\ 0
\end{array}
$$

We notice that when a column has a 5 and a 1, its digit
sum is 6. The 1 comes from the column to its right having
two 5's.

For the thousands digit of Grogg's sum to be 6, only the
larger number can have a 5 in the thousands place, but
both numbers must have a 5 in the hundreds place. For
example:

$$
\begin{array}{r}
1\ \ \ 1\ 1\ \ \ \\
5\ 5,5\ 5\ 5\\
+\ \ \ \ \ 5\ 5\ 5\\
\hline
5\ 6,1\ 1\ 0
\end{array}
$$

So, the smaller number Grogg adds is **555**, since it has
no thousands digit and a 5 in the hundreds place.

The larger number has a 5 in the thousands place, so
it is four or more digits long. For example, the larger
number could be 5,555, or 555,555,555,555.

145. To make the difference as small as possible, Lizzie's two
numbers must be as close together as possible. So, their
thousands digits must be consecutive. For example, if
the smaller number has thousands digit 5, then the larger
number must have thousands digit 6.

Additionally, we want the smaller number to be as large
as possible and the larger number to be as small as
possible.

If we make the three rightmost digits of the smaller
number as large as we can, we get $\boxed{},765$. If we make
the three rightmost digits of the larger number as small
as we can, we get $\boxed{},012$.

The remaining digits are 3 and 4, which are consecutive!
So, we fill the blanks in the thousands place to get 3,765
and 4,012. Then, we subtract to find the difference.

$$
\begin{array}{r}
9\ \ 10\ \ \ \ \\
3\ \ \overset{\scriptstyle}{10}\ \overset{\scriptstyle}{0}\ 12\\
\cancel{4},\cancel{0}\ \cancel{1}\ 2\\
-\ 3,7\ 6\ 5\\
\hline
2\ 4\ 7
\end{array}
$$

So, the smallest possible difference between Lizzie's two
numbers is **247**.

146. We first stack the numbers.

The largest possible sum of three 4-digit numbers is $9,999+9,999+9,999 = 29,997$. So, the thousands digit of the sum is 1 or 2.

```
  A , A A A
  B , B B B
+ C , C C C
A B , B B C
```

So, A is 1 or 2. First, we try A = 2.

```
  2 , 2 2 2
  B , B B B
+ C , C C C
2 B , B B C
```

In the ones column, we can only add 0 or 10 to C to get a sum with ones digit C. So, if A is 2, then 2+B is 10, and B is 8.

Since the sum of the ones column is between 10 and 19, we place an extra 1 in the tens column.

```
          1
    2 , 2 2 2
    8 , 8 8 8
+   C , C C C
2 8 , 8 8 C
```

The sum of the tens column is only correct if C is 7.

But, $2,222+8,888+7,777$ does not equal $28,887$. So, A is not 2. ✖

```
          1
    2 , 2 2 2
    8 , 8 8 8
+   7 , 7 7 7
2 8 , 8 8 7
```

Next, we try A = 1.

In the ones column, we can only add 0 or 10 to C to get a sum with ones digit C. So, if A is 1, then 1+B is 10, and B is 9.

Since the sum of the ones column is between 10 and 19, we place an extra 1 in the tens column.

```
          1
    1 , 1 1 1
    9 , 9 9 9
+   C , C C C
1 9 , 9 9 C
```

The sum of the tens column is only correct if C is 8.

This works!

So, we have A = **1**, B = **9**, and C = **8**.

```
    1   1 1
    1 , 1 1 1
    9 , 9 9 9
+   8 , 8 8 8
1 9 , 9 9 8
```

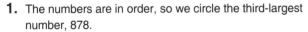
Organize   71

1. The numbers are in order, so we circle the third-largest number, 878.

777 778 787 788 877 (878) 887 888

2. We start by ordering the numbers from least to greatest. Then, we circle the third-largest number, 689.

542 588 644 646 647 (689) 690 719

3. The numbers **32** and **38** are missing from the list.

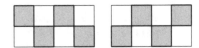

30 31 33 34 35 36 37 39 40 41 42

4. We start by ordering the numbers from least to greatest. Then, it is easy to see that **49** and **54** are missing from the list.

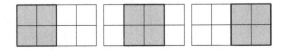

45 46 47 48 50 51 52 53 55 56 57

Counting Shapes   72-73

5. We count **8** small squares.

Then, there are **3** large squares.

So, there are a total of 8+3 = **11** squares.

6. We count **3** small triangles.

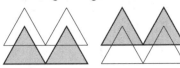

Then, there are **4** large triangles.

So, there are a total of 3+4 = **7** triangles.

7. We count **4** small triangles.

Then, there are **4** large triangles.

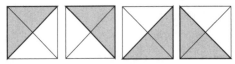

So, there are a total of 4+4 = **8** triangles.

8. We count 5 small squares and 4 large squares.

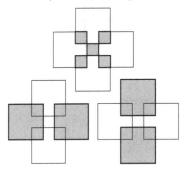

So, there are a total of 5+4 = **9** squares.

9. We count 14 small triangles and 6 large triangles.

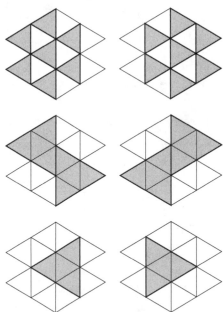

So, there are a total of 14+6 = **20** triangles.

To organize our work in the solutions below, we always try to go right before going down. At each intersection, we only go down if going right gives us a path we have already traced. You may have organized your work differently to get the same paths.

10. The three paths are shown below.

11. The four paths are shown below.

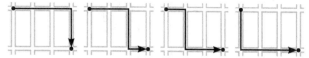

12. The four paths are shown below.

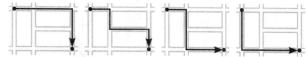

13. There are **5** possible paths, as shown below.

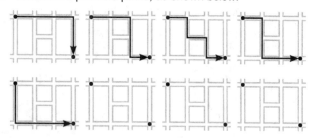

14. There are **6** possible paths, as shown below.

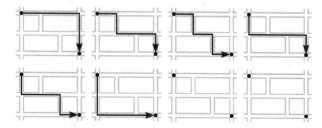

15. There are **6** possible paths, as shown below.

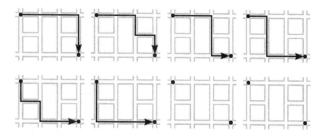

16. There are **8** possible paths, as shown below.

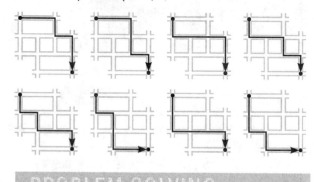

17. From the Start dot, if we go right, there is only one way to finish the path. If we go left, there is only one way to finish the path. We can't cross every square if we begin by going down. We draw both paths as shown below.

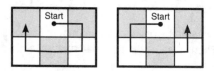

18. The three possible paths are shown below.

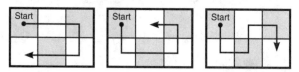

19. The three possible paths are shown below. Dots are shown on squares where we have to choose a direction.

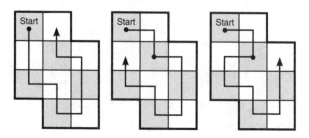

20. We organize our work to make sure we don't miss any paths. From the Start dot, we can go either down or right. If we start by going down, our next move has to be to the right, as shown.

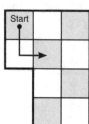

From here, we can either go up or down (going to the right makes it impossible to pass through every square). If we go up, we cross the next three squares as shown before we reach a square with more than one option.

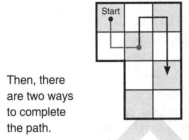

Then, there are two ways to complete the path.

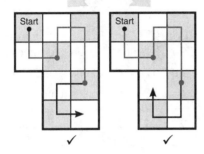

If we go down instead of up, there is only one way to complete the path.

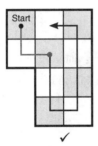

Next, we look for paths that begin by going right.

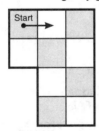

If we start by going right, there is only one way to complete the path.

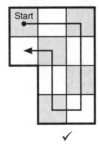

This gives us a total of **4** Checkerboard Paths, all marked with ✓'s.

21. We can start by going left or down. There are three ways to complete the path if we start by going left. Dots are shown on squares where we have to choose a direction.

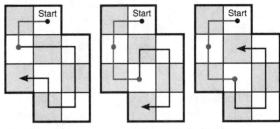

If we start by going down instead of left, there is only one way to complete the path.

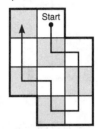

So, there are **4** Checkerboard Paths.

22. We can start by going right or down. If we start by going right, there are three ways to complete the path. Dots are shown on squares where we have to choose a direction.

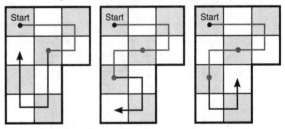

If we start by going down, there are two ways to complete the path.

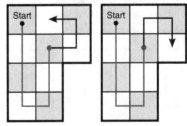

So, there are **5** Checkerboard Paths.

23. We can start by going right or down. If we start by going right, there is only one way to complete the path.

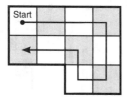

If we start by going down, there are four ways to complete the path. Dots are shown on squares where we have to choose a direction.

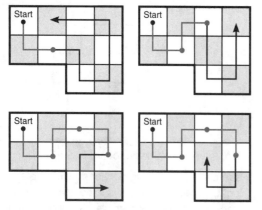

So, there are **5** Checkerboard Paths.

24. We can start by going right, left, or down. If we start by going right, there is only one way to complete the path.

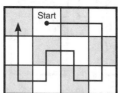

If we start by going left, there are four ways to complete the path. Dots are shown on squares where we have to choose a direction.

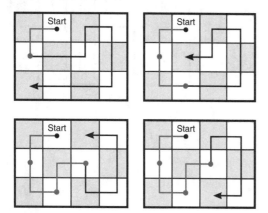

Finally, if we start by going down, there is only one way to complete the path.

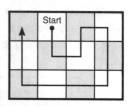

So, there are **6** Checkerboard Paths.

PROBLEM SOLVING
Colorominoes   82-85

25. The three ways to shade two squares are shown below. When shading two squares, we leave either the left, middle, or right square unshaded.

26. The shapes ▭▬ and ▬▭ can be turned to look the same. So, ▭▬ and ▬▭ are the same coloromino.

There are **2** different colorominoes: ▭▬ and ▬▭.

27. The four ways to shade one square are shown below.

28. The shapes ⌐▭ and ⌐▬ can be turned to look the same.

Similarly, the shapes ▬⌐ and ▭⌐ can be turned to look the same.

So, there are **2** different colorominoes: ⌐▭ and ⌐▬.

29. The six ways to shade two squares are shown below.

30. The shapes ▣, ▤, ▥, and ▦ can be turned to look the same.

Similarly, the shapes ▧ and ▨ can be turned to look the same.

So, there are **2** different colorominoes: ▣ and ▨.

31. The six ways to shade five squares are shown below. When shading five squares, one square is left unshaded.

In this problem, it is easier to think about which square is unshaded when organizing your work.

32. The shapes ▦ and ▦ can be turned to look the same.

Similarly, the shapes ▦ and ▦ can be turned to look the same.

Finally, the shapes ▦ and ▦ can be turned to look the same.

So, there are **3** different colorominoes: ▦, ▦, and ▦.

33. The six ways to shade two squares are shown below.

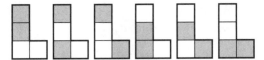

34. None of these shapes can be turned to look the same.

So, there are **6** colorominoes: 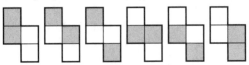 and 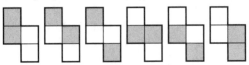.

35. The six ways to shade two squares are shown below.

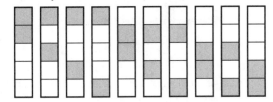

36. The shapes and can be turned to look the same.

Similarly, the shapes and can be turned to look the same.

So, there are **4** colorominoes: , , , and .

37. The ten ways to shade two squares are shown below.

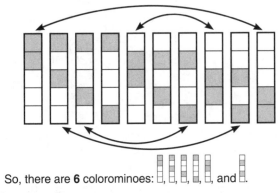

38. Below, we connect the four pairs of shapes that can be turned to look the same.

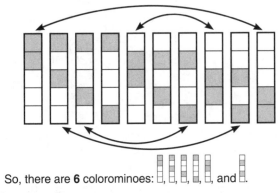

So, there are **6** colorominoes: 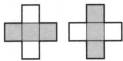, , , , , and .

39. First, we consider colorominoes that have the center square shaded.

If the center square is shaded, we need to shade two more squares. We can shade two of the "arms." The arms can be next to each other, as shown below.

These are all the same coloromino.

Or, the arms can be on opposite sides.

These are the same coloromino.

Then, we consider colorominoes that do not have the center square shaded. Then, we must shade three of the arms.

These are all the same coloromino.

So, the three different colorominoes are shown below.

You may have shaded different squares in your three colorominoes to create the same three colorominoes shown above.

— *or* —

Instead of thinking about the squares that *are* shaded, we consider squares that are *not* shaded. This is a little easier, since there are only two of them.

If the center square is unshaded, one of the arms must be unshaded.

If the center square is shaded, then two arms must be unshaded. There are two ways to leave two arms unshaded. The arms can be next to each other, or on opposite sides.

This gives us the same three colorominoes shown in the first solution.

40. There are two numbers in the 60's: 66 and 67.

There are two numbers in the 70's: 76 and 77.

In order, we have __66__ , __67__ , __76__ , __77__ .

41. The 1-digit numbers are 0 and 1.

The 2-digit numbers are 10 and 11 (0 cannot be the tens digit).

The 3-digit numbers are 100, 101, 110, and 111.

So, we fill in the blanks as shown.

__0__ , __1__ , __10__ , __11__ , __100__ , __101__ , __110__ , __111__ .

42. We list the numbers in order.

In the 200's, we have 222, 223, 232, and 233.

In the 300's, we have 322, 323, 332, and 333.

So, there are a total of **8** numbers.

43. When we arrange three 5's and one 7, we can place the 7 in the ones, tens, hundreds, or thousands place. In order, we have

<u>5,557</u> , <u>5,575</u> , <u>5,755</u> , <u>7,555</u> .

44. There are two numbers in the 200's, two numbers in the 300's, and two numbers in the 400's.

<u>234</u> , <u>243</u> , <u>324</u> , <u>342</u> , <u>423</u> , <u>432</u> .

45. We cannot start our number with 0. So, the thousands digit must be 1 or 2.

If 1 is the thousands digit, then 2 can be in the ones, tens, or hundreds place.

1,002   1,020   1,200

If 2 is the thousands digit, then 1 can be in the ones, tens, or hundreds place.

2,001   2,010   2,100

In order, we have

<u>1,002</u> , <u>1,020</u> , <u>1,200</u> , <u>2,001</u> , <u>2,010</u> , <u>2,100</u> .

46. If 1 is the thousands digit, then the other 1 can be in the hundreds, tens, or ones place.

1,199   1,919   1,991

If 9 is the thousands digit, then the other 9 can be in the hundreds, tens, or ones place.

9,911   9,191   9,119

So, there are a total of **6** four-digit numbers.

PROBLEM SOLVING

Organizing Sums   88-89

47. We start by using as many 2's as possible. Then, we can replace one, two, or three 2's with 1+1 to get the other ways.

2+2+2+1

2+2+1+1+1

2+1+1+1+1+1

1+1+1+1+1+1+1

48. We organize our work by the number of 4's in the sum. We begin with two 4's.

4+4

With one 4, there is only one way to get a sum of 8.

4+2+2

With zero 4's, we can use two 3's and one 2, or four 2's.

3+3+2

2+2+2+2

49. We organize our work by the number of 3's in the sum.

3+3+3+1

3+3+1+1+1+1

3+1+1+1+1+1+1+1

1+1+1+1+1+1+1+1+1+1

There are **4** ways to get a sum of 10 using 1's and 3's.

50. We organize our work by the number of 5's in the sum, then by the number of 3's.

There are two ways to use one 5.

5+3

5+1+1+1

There are three ways to use zero 5's. We organize our work by the number of 3's.

3+3+1+1

3+1+1+1+1+1

1+1+1+1+1+1+1+1

This gives us a total of **5** ways to get a sum of 8 using 1's, 3's, and 5's.

51. We organize our work by the number of 3's in the sum, then by the number of 2's.

There is one way to use two 3's.

3+3

There are two ways to use one 3.

3+2+1

3+1+1+1

There are four ways to use zero 3's. We organize our work by the number of 2's.

2+2+2

2+2+1+1

2+1+1+1+1

1+1+1+1+1+1

This gives us a total of **7** ways to get a sum of 6 using 1's, 2's, and 3's.

PROBLEM SOLVING

Patterns   90-91

52. We add 2 to each number to get the next number in the pattern.

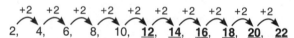

2,   4,   6,   8,   10, **12**, **14**, **16**, **18**, **20**, **22**

53. We add 2 to each number to get the next number in the pattern.

21,   23,   25,   27,   29, **31**, **33**, **35**, **37**, **39**, **41**

54. We add 10 to each number to get the next number in the pattern.

19,   29,   39,   49, **59**, **69**, **79**, **89**, **99**, **109**

55. We double each number to get the next number in the pattern.

1,   2,   4,   8,   16, **32**, **64**, **128**, **256**, **512**

56. We add 3 to each number to get the next number in the pattern.

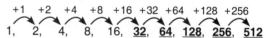

18,   21,   24,   27,   30, **33**, **36**, **39**, **42**, **45**

57. We add 1, then 2, then 3, and continue to add one more than we added to the previous number.

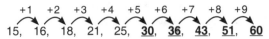

15, 16, 18, 21, 25, **30**, **36**, **43**, **51**, **60**

58. We add 11 to each number to get the next number in the pattern.

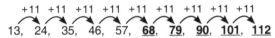

13, 24, 35, 46, 57, **68**, **79**, **90**, **101**, **112**

59. We add 2, then subtract 1, then add 2, then subtract 1, and continue this way to get each number in the pattern.

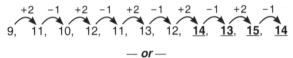

9, 11, 10, 12, 11, 13, 12, **14**, **13**, **15**, **14**

— *or* —

We get the same answer if we complete the pattern as shown below.

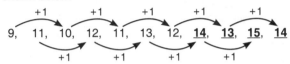

60. We double each number to get the next number in the pattern.

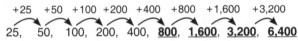

25, 50, 100, 200, 400, **800**, **1,600**, **3,200**, **6,400**

61. We add 1, then 2, then 1, then 2, and continue this way to get each number in the pattern.

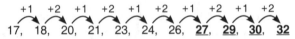

17, 18, 20, 21, 23, 24, 26, **27**, **29**, **30**, **32**

62. We add 2, then 4, then 6, then 8, and continue to add two more to each number than we added previously.

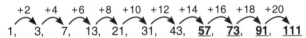

1, 3, 7, 13, 21, 31, 43, **57**, **73**, **91**, **111**

63. To get the next number in the pattern, we add the two numbers that come before it.

So, to get the number that comes after 16 and 26, we add 16+26 to get 42.

After 26 and 42 is 26+42 = 68.
After 42 and 68 is 42+68 = 110.
After 68 and 110 is 68+110 = 178.

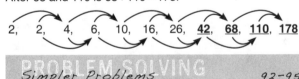

2, 2, 4, 6, 10, 16, 26, **42**, **68**, **110**, **178**

PROBLEM SOLVING
Simpler Problems   92-95

64. Lizzie reads pages 6, 7, and 8, for a total of **3** pages.

Careful! We can't just subtract 8−6 to get the number of pages Lizzie reads.

65. Lizzie reads pages 25, 26, 27, 28, 29, and 30, for a total of **6** pages.

Careful! We can't just subtract 30−25 to get the number of pages Lizzie reads.

66. This seems like it might be an easy subtraction problem. But, in the previous two problems, subtracting the first page from the last page does not give us the correct number of pages read.

Reading pages 6 through 8, Lizzie doesn't read 8−6 = 2 pages, she reads 3 pages.

Reading pages 25 through 30, Lizzie doesn't read 30−25 = 5 pages, she reads 6 pages.

The number of pages Lizzie reads seems to always be 1 more than the number you get when you subtract the first page from the last page.

So, it appears that to find out how many pages Lizzie reads from page 98 to page 152, we subtract 152−98, then add 1.

When you find a pattern, it's always good to figure out why it works. If Lizzie had read pages 1 through 152, that would be 152 pages. If we subtract 98 from this, we've taken away all 98 pages from 1 to 98. But, Lizzie *reads* page 98, so we need to add that one back.

So, Lizzie reads a total of 152−98+1 = 54+1 = **55** pages.

67. We draw a line that connects each dot to the other three dots to get a total of **6** lines.

68. We draw a line that connects each dot to the other four dots to get a total of **10** lines.

Notice that this diagram looks a lot like the previous diagram, but with one extra dot that is connected by 4 lines to the previous dots.

69. We draw a line that connects each dot to the other five dots to get a total of **15** lines.

Notice that this diagram looks a lot like the previous diagram, but with one extra dot that is connected by 5 lines to the previous dots.

70. We have diagrams for circles with 3, 4, 5, and 6 dots.

To connect 3 dots takes 3 lines.
To connect 4 dots takes 6 lines.
To connect 5 dots takes 10 lines.
To connect 6 dots takes 15 lines.

We can look at the pattern of the numbers of lines for 3, 4, 5, and 6 dots.

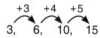

Each new dot must be connected by a line to all of the old dots that were already there.

So, the 4th dot adds 3 new lines.
The 5th dot adds 4 new lines.
The 6th dot adds 5 new lines, and
the 7th dot will add 6 new lines.

So, to connect 7 dots takes $15+6=\textbf{21}$ lines.

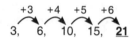

71. Every tripticorn has 1 more leg than it has horns. So, if there is only 1 tripticorn, it will have 1 more leg than horns.

If there are 2 tripticorns, there will be $1+1=2$ more legs than horns.

If there are 3 tripticorns, there will be $1+1+1=3$ more legs than horns.

The pattern continues, with each tripticorn adding 1 more leg than it adds horns. So, if there are 26 tripticorns, there will be **26** more legs than horns in the barn.

72. Cutting one carrot into 2 pieces takes just 1 cut.

Cutting one carrot into 3 pieces takes 2 cuts.

Cutting one carrot into 4 pieces takes 3 cuts.

The pattern continues. The number of cuts is always one less than the number of pieces. So, to split one carrot into 50 pieces takes **49** cuts.

73. Folding a sheet of paper in half 1 time splits it into 2 rectangles.

Folding a sheet in half 2 times splits both of the first two rectangles in half, creating $2+2=4$ rectangles.

Folding a sheet in half 3 times splits all four rectangles in half, creating $4+4=8$ rectangles.

Folding a sheet in half 4 times splits all eight rectangles in half, creating $8+8=16$ rectangles.

We keep doubling the number of rectangles to see that 5 folds creates $16+16=32$ rectangles, 6 folds creates $32+32=64$ rectangles, and 7 folds creates $64+64=\textbf{128}$ rectangles.

If you are trying to do this yourself, you may find it's very hard to fold a piece of paper in half 7 times, and nearly impossible to fold a piece of paper in half 8 times!

74. Adding 10 copies of 1 gives us 10.
Adding 10 copies of 2 gives us 20.
Adding 10 copies of 3 gives us 30.
Adding 10 copies of 4 gives us 40.

The pattern continues. To get the sum of 10 copies of a number, we just write the digits of the number followed by a zero (in the ones place).

So, the sum of 10 copies of 74 is **740**.

Adding 10 copies of 74 gives the same sum as adding 74 copies of 10 (we will learn more about this when we get to multiplication in Beast Academy 3B). 74 tens is 740.

75. A stack of 2 cups is 5 inches tall.
A stack of 3 cups is 6 inches tall.
A stack of 4 cups is 7 inches tall.

The pattern continues, with each cup adding 1 more inch to the height of the stack. The height of the stack is always 3 inches more than the number of cups. So, a stack of 20 cups is $20+3=\textbf{23}$ inches tall.

76. To make 1 triangle takes 3 toothpicks.
To make 2 triangles takes 5 toothpicks.
To make 3 triangles takes 7 toothpicks.

The pattern continues, with each triangle adding 2 more toothpicks. So, for 25 triangles, we have 3 toothpicks for the first triangle plus 2 toothpicks for each of the other 24 triangles. Adding 24 twos gives us 48. So, it takes a total of $3+48=\textbf{51}$ toothpicks.

— or —

Looking at the pattern, the number of toothpicks is always 1 more than double the number of triangles. So, the number of toothpicks needed to make 25 triangles is $25+25+1=\textbf{51}$ toothpicks.

PROBLEM SOLVING
Eliminating Choices   *96-97*

77. $45,561+19,871$ is about $45,000+20,000=65,000$. We can eliminate all of the answer choices that are not close to 65,000. We circle the only answer choice that makes sense, 65,432.

~~432~~   ~~5,432~~   (65,432)   ~~765,432~~   ~~8,785,432~~

78. When we add the ones digits of 148,32<u>7</u> and 98,53<u>7</u>, we get $7+7=1\underline{4}$. So, the ones digit of the sum is 4. The only answer choice with ones digit 4 is 246,864.

~~246,863~~   (246,864)   ~~246,865~~   ~~246,866~~   ~~246,867~~

79. Adding just the thousands in $14,296+13,297+12,298$, we get $14,000+13,000+12,000=39,000$.

So, $14,296+13,297+12,298$ is more than 39,000. But, the sum is much less than 83,198. We circle the only answer choice that makes sense, 39,891.

~~27,981~~   ~~38,189~~   (39,891)   ~~83,198~~   ~~93,891~~

80. Adding 6's will always give an even result. So, we circle the only even answer choice, 738.

~~731~~   ~~733~~   ~~735~~   ~~737~~   (738)

81. We learned in Problem 74 that adding 10 copies of a number always gives a result that ends in 0. So, no matter how many crayons are in each box, the total number of crayons in 10 boxes ends in 0. We circle the only answer choice that ends in 0, which is 320.

~~317~~ ~~318~~ ~~319~~ (320) ~~321~~

82. Since every stardvark has 5 snouts, we can find the total number of snouts by counting by 5's. When we count by 5's, every number ends in 0 or 5: 5, 10, 15, 20, 25, 30, and so on. So, the total number of snouts must end in 0 or 5. We circle the only answer choice that ends in 0 or 5, which is 215.

~~211~~ ~~212~~ ~~213~~ ~~214~~ (215)

83. Adding even numbers always gives an even result. So, we can eliminate all of the odd answer choices.

~~24,930,561~~ 24,930,562 ~~24,930,563~~ 34,930,562 34,930,564

The six numbers we are adding only include one number that is more than 1 million (24,681,012). So, the sum is less than the two answer choices that are more than 34 million. We eliminate these and circle the only remaining answer choice that is less than 34 million, 24,930,562.

~~24,930,561~~ (24,930,562) ~~24,930,563~~ ~~34,930,562~~ ~~34,930,564~~

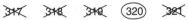

PROBLEM SOLVING

River Crossings   98-99

84. We can't send the 30-pound monster across first, since the monster will just have to paddle back across to bring the boat back. So, we start by sending all three 10-pound monsters across.

We send one 10-pound monster back.

Now, we can send the 30-pound monster across.

We send one 10-pound monster back.

Finally, the two 10-pound monsters cross together.

The boat crosses the river 5 times.

We could have also done this by sending just two 10-pound monsters across in the first trip, which would leave three 10-pound monsters to cross in the fifth trip, but **5** is the smallest number of trips possible.

85. We can deliver the monsters across with the steps shown below.

The boat crosses the river 5 times.

We could also switch the 20- and 30-pound monsters above, but **5** is the smallest number of trips possible.

86. We send two 10-pound monsters across. One comes back. Then, we send a 20-pound monster across, and the second 10-pound monster returns.

We repeat the same steps to get the second 20-pound monster across.

Finally, the two 10-pound monsters cross together.

The boat must cross the river a total of **9** times.

87. The fox and corn are the only two items that can be left alone together. So, on her first trip, the farmer can only take the hen across.

She leaves the hen alone and returns with the boat. (Since the farmer always has to be in the boat, we do not need a slip of paper to show her.)

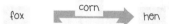

She can now choose either the fox or the corn to take across. We try the corn.

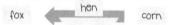

She can't leave the corn and the hen alone, so she has to come back with one or the other. Returning with the corn would undo her previous trip, so she returns with the hen.

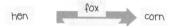

She can't leave the fox and the hen alone. Taking the hen across would undo her previous trip, so she crosses with the fox.

Now, she can leave the fox and the corn alone. She crosses to get the hen, then brings it across.

For the third trip shown above, the farmer could have chosen to take the fox instead of the corn. All of the steps that follow would have been the same, but with the fox and corn switched.

Either way, the boat must cross the river a total of **7** times.

88. We can switch each pawn with the opposite-colored pawn across from it as shown below.

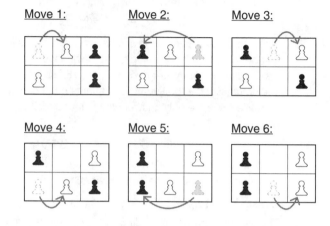

You may have found another sequence of moves, but **6** is the smallest number of moves needed.

89. We can switch each pawn with the opposite-colored pawn across from it as shown below.

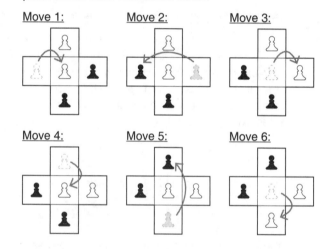

You may have found another sequence of moves, but **6** is the smallest number of moves needed.

90. We can switch the pawns with the moves shown below.

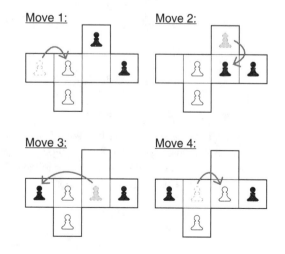

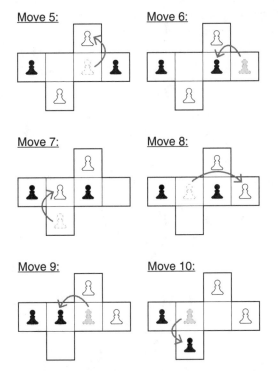

Move 5:

Move 6:

Move 7:

Move 8:

Move 9:

Move 10:

You may have found another sequence of moves, but **10** is the smallest number of moves needed.

91. We can switch the pawns with the moves shown below.

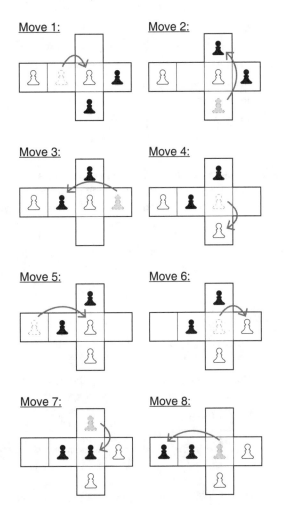

Move 1:

Move 2:

Move 3:

Move 4:

Move 5:

Move 6:

Move 7:

Move 8:

You may have found another sequence of moves, but **8** is the smallest number of moves needed.

92. We can switch the pawns with the moves shown below.

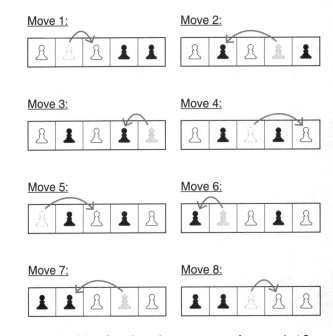

Move 1:

Move 2:

Move 3:

Move 4:

Move 5:

Move 6:

Move 7:

Move 8:

You may have found another sequence of moves, but **8** is the smallest number of moves needed.

93. We can switch the pawns with the moves shown below.

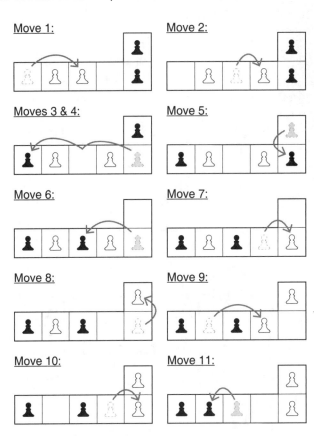

Move 1:

Move 2:

Moves 3 & 4:

Move 5:

Move 6:

Move 7:

Move 8:

Move 9:

Move 10:

Move 11:

You may have found another sequence of moves, but **11** is the smallest number of moves needed.

94. We can switch the pawns with the moves shown below.

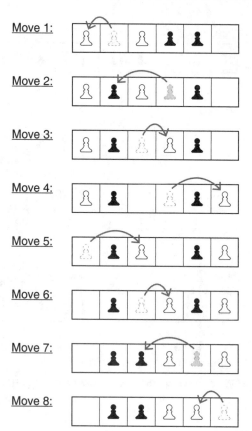

You may have found another sequence of moves, but **8** is the smallest number of moves needed.

95. We can switch the pawns with the moves shown below.

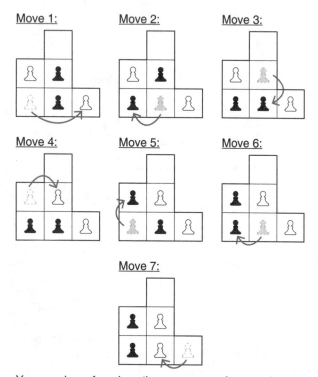

You may have found another sequence of moves, but **7** is the smallest number of moves needed.

96. We can switch the pawns with the moves shown below.

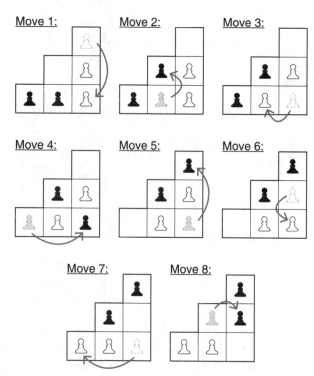

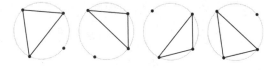

You may have found another sequence of moves, but **8** is the smallest number of moves needed.

97. We organize our work by the hundreds digit.

In the 700's, we have 703, 712, 721, and 730.
In the 800's, we have 802, 811, and 820.
In the 900's, we have 901 and 910.

So, there are **9** numbers between 700 and 1,000 whose digits add up to 10.

98. To organize our work, we can choose one of the four dots to leave out, connecting the other three. Since there are four dots that can be left out, there are **4** ways to connect three of the four dots, as shown below.

99. To get from two 2's to three 3's, we add 5.
To get from three 3's to four 4's, we add 7.
To get from four 4's to five 5's, we add 9.

We continue the pattern, adding the next odd number to get from one empty box to the next.

two 2's	three 3's	four 4's	five 5's	six 6's	seven 7's	eight 8's	nine 9's	ten 10's	eleven 11's
4	9	16	25	36	49	64	81	100	121

+5 +7 +9 +11 +13 +15 +17 +19 +21

The numbers in this table are called perfect squares. You'll learn more about them in Beast Academy 3B.

100. We organize our work by how many "blocks" the mouse goes to the right before turning down.

If the mouse starts by going all the way to the right, there is only one way to reach the cheese.

If the mouse goes two blocks to the right before turning down, there are two ways to reach the cheese.

If the mouse goes one block to the right before turning down, there are three ways to reach the cheese.

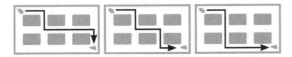

If the mouse starts by going down instead of right, there are four ways to reach the cheese.

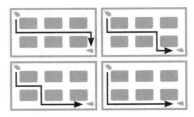

So, there are a total of $1+2+3+4 = \mathbf{10}$ ways for the mouse to reach the cheese.

You may have noticed that this is the same as the example given for the Taxi Path puzzles.

101. Below are the ten ways to shade three triangles.

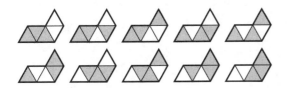

Organizing the ten ways shown above can be tricky.

We can organize the ten ways to shade three triangles by choosing which triangles to leave *unshaded*, starting with the triangles farthest to the left.

There are 4 ways that leave the leftmost triangle unshaded, 3 ways that leave the second triangle unshaded (but shade the first), 2 ways that leave the third triangle unshaded (but shade the first two) and 1 way that leaves the fourth triangle unshaded (but shades the first three).

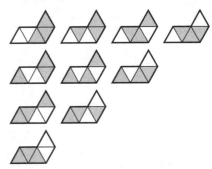

102. We organize our work by how many dimes (10's) we use, then by how many nickels (5's) we use. We write each way to get 25 cents as a sum of 10's, 5's, and 1's.

There are 2 ways that include 2 dimes.

$$10+10+5$$
$$10+10+1+1+1+1+1$$

Next, there are 4 ways that include 1 dime.

$$10+5+5+5$$
$$10+5+5+1+1+1+1+1$$
$$10+5+\text{ten 1's}$$
$$10+\text{fifteen 1's}$$

Finally, there are 6 ways that include 0 dimes.

$$5+5+5+5+5$$
$$5+5+5+5+1+1+1+1+1$$
$$5+5+5+\text{ten 1's}$$
$$5+5+\text{fifteen 1's}$$
$$5+\text{twenty 1's}$$
$$\text{twenty-five 1's.}$$

So, there are a total of $2+4+6 = \mathbf{12}$ ways to use pennies, nickels, and dimes to make 25 cents.

103. We can switch the pawns with the moves shown below.

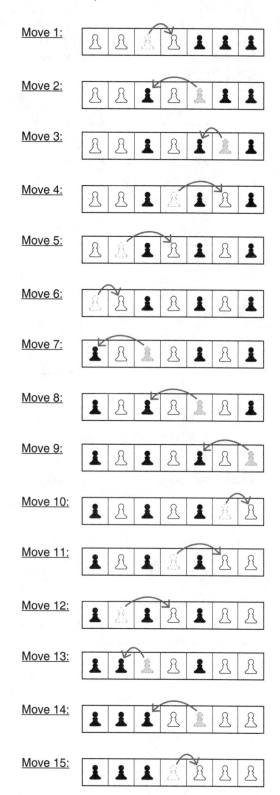

Move 1:

Move 2:

Move 3:

Move 4:

Move 5:

Move 6:

Move 7:

Move 8:

Move 9:

Move 10:

Move 11:

Move 12:

Move 13:

Move 14:

Move 15:

You may have found another sequence of moves, but **15** is the smallest number of moves needed.

Don't worry if you didn't get this one. Even the authors of this book had a hard time coming up with the smallest number of moves!

For additional books, printables, and more, visit
BeastAcademy.com